Advances in Experimental Medicine and Biology

Proteomics, Metabolomics, Interactomics and Systems Biology

Advances in Experimental Medicine and Biology presents multidisciplinary and dynamic findings in the broad fields of experimental medicine and biology. The wide variety in topics it presents offers readers multiple perspectives on a variety of disciplines including neuroscience, microbiology, immunology, biochemistry, biomedical engineering and cancer research. *Advances in Experimental Medicine and Biology* has been publishing exceptional works in the field for over 30 years and is indexed in Medline, Scopus, EMBASE, BIOSIS, Biological Abstracts, CSA, Biological Sciences and Living Resources (ASFA-1), and Biological Sciences. The series also provides scientists with up-to-date information on emerging topics and techniques. 2018 Impact Factor: 2.126.

Content of this series is reviewed.

This series of volumes focuses on concepts, techniques and recent advances in the field of proteomics, interactomics, metabolomics and systems biology. Recent advances in various 'omics' technologies enable quantitative monitoring of myriad various biological molecules in a high-throughput manner, and allow determination of their variation between different biological states on a genomic scale. Now that the sequencing of various genomes, from prokaryotes to humans, has provided the list and linear sequence of proteins and RNA that build living organisms, defining the complete set of interactions that sustain life constitutes one of the key challenges of the postgenomic era. This series is intended to cover experimental approaches for defining protein-protein, protein-RNA, protein-DNA and protein-lipid interactions; as well as theoretical approaches dealing with data analysis, integration and modeling and ethical issues.

More information about this series at http://www.springer.com/series/5584

Hugo Miguel Baptista Carreira dos Santos
Editor

Translational Urinomics

Springer

Editor
Hugo Miguel Baptista Carreira dos Santos
LAQV-REQUIMTE - Department of Chemistry
NOVA School of Science and Technology
NOVA University of Lisbon
Caparica, Portugal

Proteomics, Metabolomics, Interactomics and Systems Biology
ISSN 0065-2598 ISSN 2214-8019 (electronic)
Advances in Experimental Medicine and Biology
ISBN 978-3-030-63907-5 ISBN 978-3-030-63908-2 (eBook)
https://doi.org/10.1007/978-3-030-63908-2

This Springer imprint is published by the registered company Springer Nature Switzerland AG
The registered company address is: Gewerbestrasse 11, 6330 Cham, Switzerland

Preface

Urine as a reservoir of health information can be traced back in history to the early days of human writing records [1, 2]. Thus, in modern time, *Schistosoma haematobium* infestation was first described in the nineteenth century by Theodore Bilharz, and yet it was mentioned in ancient Egypt, in Ebers papyrus dated 1550 BC [3–5]. The same papyrus also mentions the use of yellow ochre as a remedy for urological complaints. In ancient India, the Atharva Veda text talks for the first time about urological diseases and urinary retention. Later in the ninth century BC, the writing of the Indian physician Charaka includes a volume devoted to urinalysis and clinical interpretations based upon color consistency, turbidity, thickness, presence of blood, semen, pus, and fat in the urine. The text also offers diagnosis based on symptoms, and links urinary retention to dietary and alcoholic indiscretions [6].

In ancient China, the use of herbs as part of a medical culture can be traced back to circa 3000 BC. However, the first medical system developed and organized was first realized during the Zhou dynasty (221 BC to 100 BC). Herbal medicines to deal with urological problems are mentioned here [7].

In ancient Mesopotamia (Babylon), medicine was well developed. There are writing registers reporting more that 250 medicinal plants, 120 mineral extracts and 180 other substances. For instance, incontinence was treated, introducing medications into the urethra using a bronze catheter [8].

In ancient Greece, Hippocrates (496 BC–399 BC) made several statements that have remained unaltered till today, namely: urine is made in the kidneys and flows into the bladder; blood in the urine comes from the kidneys or bladder; strangury, dysuria, and anuria are different symptoms; and urinary symptoms are caused by diseases of the bladder, kidneys or urethra [9]. Finally, in ancient Rome, medicine evolved from the techniques employed by ancient Greek and Egyptian physicians. The work of Aulus Celsus *De Medicina* describes perineal lithotomy for bladder stone [10].

As history unfolds and technology progresses, the use of urine as a liquid biopsy to decipher the healthy status of a patient is gaining momentum. The advent of methods that allows analyzing a wide array of analytes, such as proteins, metabolites, DNA, and urinary exosomes has granted possibilities never envisioned before in medicine. Thus, it is becoming clear that many diseases can be reflected in one way or another in the molecular features present in urine. In our group, for instance, we are now developing methods for monitoring response to therapy using urine, thus providing near real-time information about a patient's response to treatment.

In this book, we present an overview of urine analysis using proteomics, metabolomics, and immunoreactivity assays as well as other technology-driven techniques for the diagnosis and prognosis of diseases.

I am in debt to my fellow contributors, who accepted my welcome to write a chapter for this book. Also, to the editorial staff at Springer-Nature for allowing me to launch the book and for giving me the necessary advice at all time. Finally, a word of gratitude to the members of my team and the Bioscope Research Group, for their patience during the organization of this book.

Caparica, Portugal — Hugo Miguel Baptista Carreira dos Santos
Pittsburgh, PA, USA

References

1. Armstrong, J.A.: Urine in Western culture: a brief history. Kidney Int. **71**, 384–387 (2007). https://doi.org/10.1038/sj.ki.5002057.
2. Lodeiro, C., Capelo, J.L., Santos, H.M. The aurum sample. Proteomics Clin. Appl. **9**, 449–452 (2015). https://doi.org/10.1002/prca.201570033.
3. Di Bella, S., Riccardi, N., Giacobbe, D.R., Luzzati, R. History of schistosomiasis (bilharziasis) in humans: from Egyptian medical papyri to molecular biology on mummies. Pathog Glob. Health. **112**(5), 268–273 (2018 July). https://doi.org/10.1080/20477724.2018.1495357.
4. https://www.baus.org.uk/museum/76/ancient_egypt. Last time accessed 8 Oct 2020.
5. https://en.wikipedia.org/wiki/Ebers_Papyrus. Last time accessed 8 Oct 2020.
6. https://www.baus.org.uk/museum/90/ancient_india. Last time accessed 8 Oct 2020.
7. https://www.baus.org.uk/museum/84/ancient_china. Last time accessed 8 Oct 2020.
8. https://www.baus.org.uk/museum/86/mesopotamia. Last time accessed 8 Oct 2020.
9. https://www.baus.org.uk/museum/91/ancient_greece. Last time accessed 8 Oct 2020.
10. White, W.I. A new look at the role of urinalysis in the history of diagnostic medicine. Clin. Chem. **37**(1), 119–125. https://doi.org/10.1093/clinchem/37.1.119.

Acknowledgments

PROTEOMASS Scientific Society is acknowledged for the funding provided to the Laboratory for Biological Mass Spectrometry – Isabel Moura. H. M. S. acknowledges the funding provided by the Associate Laboratory for Green Chemistry LAQV which is financed by national funds from FCT/MCTES (UIDB/50006/2020 and UIDP/50006/2020) and the funding provided by FCT/MCTES within the framework of FCT Investigator Program (IF/00007/2015). H. M. S. acknowledges Professor José Luís Capelo for the fruitful discussions and valuable comments about the content of Translational Urinomics book.

Contents

Contributors

Saima Ahmed Departments of Pathology, Boston Children's Hospital and Harvard Medical School, Boston, MA, USA

Alexandra M. M. Antunes Centro de Química Estrutural (CQE), Instituto Superior Técnico, Universidade de Lisboa, Lisbon, Portugal

Jéssica Freitas de Araújo Encinas Division of Clinical Analysis, Centro Universitário Saúde ABC, Santo André, Brazil

Karina Barreiro Institute of Molecular Medicine Finland, University of Helsinki, Helsinki, Finland

Otto Bekers Unit of Clinical Chemistry, Central Diagnostic Laboratory, Maastricht University Medical Center, Maastricht, The Netherlands

CARIM, School for Cardiovascular Diseases, Faculty of Health, Medicine and Life Sciences (FHML) of Maastricht University (UM), Maastricht University Medical Center, Maastricht, The Netherlands

Tue B. Bennike Departments of Pathology, Boston Children's Hospital and Harvard Medical School, Boston, MA, USA

Department of Health Science and Technology, Aalborg University, Aalborg, Denmark

Precision Vaccines Program, Boston Children's Hospital, Boston, MA, USA

Irina Bobkova Tareev Clinic, Department of Nephrology, Sechenov First Moscow State Medical University, Moscow, Russia

Douwe de Boer Unit of Clinical Chemistry, Central Diagnostic Laboratory, Maastricht University Medical Center, Maastricht, The Netherlands

NUTRIM, School of Nutrition and Translational Research in Metabolism, Medicine and Life Sciences (FHML) of Maastricht University (UM), Maastricht University Medical Center, Maastricht, The Netherlands

Natalia Chebotareva Tareev Clinic, Department of Nephrology, Sechenov First Moscow State Medical University, Moscow, Russia

Marlete Brum Cleff Group of Research, Teaching and Extension in Natural Products in the Veterinary Medical Clinic (FITOPEET), Department of Veterinary Clinics, Veterinary College, Federal University of Pelotas, Pelotas, RS, Brazil

Nuno R. Coelho CEDOC, Chronic Diseases Research Centre, NOVA Medical School|Faculdade de Ciências Médicas, Universidade NOVA de Lisboa, Lisbon, Portugal

M. João Correia CEDOC, Chronic Diseases Research Centre, NOVA Medical School|Faculdade de Ciências Médicas, Universidade NOVA de Lisboa, Lisbon, Portugal

Beatriz da Costa Aguiar Alves Division of Clinical Analysis, Centro Universitário Saúde ABC, Santo André, Brazil

William P. T. M. van Doorn Unit of Clinical Chemistry, Central Diagnostic Laboratory, Maastricht University Medical Center, Maastricht, The Netherlands

Benoit Fatou Departments of Pathology, Boston Children's Hospital and Harvard Medical School, Boston, MA, USA

Precision Vaccines Program, Boston Children's Hospital, Boston, MA, USA

Fernando Luiz Affonso Fonseca Division of Clinical Analysis, Centro Universitário Saúde ABC, Santo André, Brazil

Pharmaceutical Science Department, Universidade Federal de São Paulo/ UNIFESP – Diadema, Butantã, São Paulo, Brazil

Youhe Gao Department of Biochemistry and Molecular Biology, Beijing Key Laboratory of Gene Engineering Drug and Biotechnology, Beijing Normal University, Beijing, China

Clara Gonçalves-Dias CEDOC, Chronic Diseases Research Centre, NOVA Medical School|Faculdade de Ciências Médicas, Universidade NOVA de Lisboa, Lisbon, Portugal

Harry Holthofer Institute of Molecular Medicine Finland, University of Helsinki, Helsinki, Finland

Department of Medicine, University Medical Center Hamburg-Eppendorf, Hamburg, Germany

Tobias B. Huber Department of Medicine, University Medical Center Hamburg-Eppendorf, Hamburg, Germany

Feng Jin Departments of Surgery, Cedars-Sinai Medical Center, Los Angeles, CA, USA

Jayoung Kim Departments of Surgery, Cedars-Sinai Medical Center, Los Angeles, CA, USA

Samuel Oschin Comprehensive Cancer Institute, Cedars-Sinai Medical Center, Los Angeles, CA, USA

Departments of Surgery and Biomedical Sciences, Cedars-Sinai Medical Center, Los Angeles, CA, USA

University of California, Los Angeles, CA, USA

Department of Urology, Ga Cheon University College of Medicine, Incheon, Republic of Korea

Lidia Lysenko Tareev Clinic, Department of Nephrology, Sechenov First Moscow State Medical University, Moscow, Russia

Carolina Silveira Mascarenhas Wild Animal Parasitology Laboratory (LAPASIL), Microbiology and Parasitology Department, Institute of Biology, Federal University of Pelotas, Pelotas, RS, Brazil

Nilesh M. Mehta Department of Anesthesiology, Critical Care and Pain Medicine, Boston Children's Hospital, Harvard Medical School, Boston, MA, USA

Alma M. A. Mingels Unit of Clinical Chemistry, Central Diagnostic Laboratory, Maastricht University Medical Center, Maastricht, The Netherlands

CARIM, School for Cardiovascular Diseases, Faculty of Health, Medicine and Life Sciences (FHML) of Maastricht University (UM), Maastricht University Medical Center, Maastricht, The Netherlands

Sergey Moiseev Tareev Clinic, Department of Nephrology, Sechenov First Moscow State Medical University, Moscow, Russia

N. K. Mondal Fluorosis Foundation of India (Camping & Working for JICA Funded 'Fluorosis Mitigation Programme' in Nagaur District, Rajasthan, India), New Delhi, India

Emília C. Monteiro CEDOC, Chronic Diseases Research Centre, NOVA Medical School|Faculdade de Ciências Médicas, Universidade NOVA de Lisboa, Lisbon, Portugal

Judit Morello CEDOC, Chronic Diseases Research Centre, NOVA Medical School|Faculdade de Ciências Médicas, Universidade NOVA de Lisboa, Lisbon, Portugal

Gertrud Müller Wild Animal Parasitology Laboratory (LAPASIL), Microbiology and Parasitology Department, Institute of Biology, Federal University of Pelotas, Pelotas, RS, Brazil

Neif Murad Cardiology Department, Centro Universitário Saúde ABC, Santo André, Brazil

Sofia A. Pereira CEDOC, Chronic Diseases Research Centre, NOVA Medical School|Faculdade de Ciências Médicas, Universidade NOVA de Lisboa, Lisbon, Portugal

Soliane Carra Perera Group of Research, Teaching and Extension in Natural Products in the Veterinary Medical Clinic (FITOPEET), Department of Veterinary Clinics, Veterinary College, Federal University of Pelotas, Pelotas, RS, Brazil

Matheus Moreira Perez Division of Clinical Analysis, Centro Universitário Saúde ABC, Santo André, Brazil

Joyce Regina Raimundo Division of Clinical Analysis, Centro Universitário Saúde ABC, Santo André, Brazil

Hugo Miguel Baptista Carreira dos Santos BIOSCOPE Research Group, LAQV- REQUIMTE, NOVA School of Science and Technology, FCT NOVA, Universidade NOVA de Lisboa, Portugal

PROTEOMASS Scientific Society, Madan Park, Caparica, Portugal

Department of Pathology, University of Pittsburgh Medical Center, Pittsburgh, PA, United States

Catarina O. Sequeira CEDOC, Chronic Diseases Research Centre, NOVA Medical School|Faculdade de Ciências Médicas, Universidade NOVA de Lisboa, Lisbon, Portugal

Muhammad Shahid Departments of Surgery, Cedars-Sinai Medical Center, Los Angeles, CA, USA

Josaine Cristina da Silva Rappeti Project Dioctophyme renale in Dogs and Cats (PRODIC), Department of Veterinary Clinics, Veterinary College, Federal University of Pelotas, Pelotas, RS, Brazil

Karina Soto CEDOC, Chronic Diseases Research Centre, NOVA Medical School|Faculdade de Ciências Médicas, Universidade NOVA de Lisboa, Lisbon, Portugal

Hospital Prof. Dr. Fernando da Fonseca, EPE, Amadora, Portugal

Hanno Steen Departments of Pathology, Boston Children's Hospital and Harvard Medical School, Boston, MA, USA

Precision Vaccines Program, Boston Children's Hospital, Boston, MA, USA

Alexander S. Streng Unit of Clinical Chemistry, Central Diagnostic Laboratory, Maastricht University Medical Center, Maastricht, The Netherlands

Glaucia Luciano da Veiga Division of Clinical Analysis, Centro Universitário Saúde ABC, Santo André, Brazil

João B. Vicente Instituto de Tecnologia Química e Biológica, Universidade Nova de Lisboa, Oeiras, Portugal

Wim H. M. Vroemen Unit of Clinical Chemistry, Central Diagnostic Laboratory, Maastricht University Medical Center, Maastricht, The Netherlands

Will K. W. H. Wodzig Unit of Clinical Chemistry, Central Diagnostic Laboratory, Maastricht University Medical Center, Maastricht, The Netherlands

NUTRIM, School of Nutrition and Translational Research in Metabolism, Medicine and Life Sciences (FHML) of Maastricht University (UM), Maastricht University Medical Center, Maastricht, The Netherlands

About the Editor

Dr. Hugo Miguel Baptista Carreira dos Santos (H-index: 22) graduated in Applied Chemistry from NOVA University of Lisbon and completed a PhD in Biochemistry from the same university in 2010. During his PhD, he stayed six months at the Turku Centre for Biotechnology (Finland) working with state-of-the-art MS instrumentation for biomedical research with Professor Garry Corthals. H. M. Santos took up a post-doc at the University of Vigo (December 2010 to March 2011) followed by a move to the Institute of Biomedicine and Biotechnology (Barcelona, Spain, April 2011 to March 2012) to advance biomedical applications of mass spectrometry and translational research in the group of Professor Xavier Avilés. In 2011, H. M. Santos moved to FCT NOVA to continue his research in Biological Mass Spectrometry. Currently, he is Assistant Researcher – FCT Investigator Program at LAQV-REQUIMTE FCT NOVA (Portugal) and Adjunct Assistant Professor at the University of Pittsburgh School of Medicine. He published 110 articles in international peer-reviewed journals, which have received more than 2100 citations (Google Scholar). He is a member of the Royal Society of Chemistry. His scientific interests are focused on (1) Identification of molecules involved in complex biological processes, characterizing their structure, and monitoring how their abundance may change during these processes, in order to gain insights into the underlying molecular mechanisms; (2) nano-proteomics and nano-medicine; (3) mass spectrometry analysis of metal complexes, bioinorganic and supramolecular systems; (4) precision biochemistry and personalized medicine; and (5) biomarker development.

Sample Preparation for High-Throughput Urine Proteomics Using 96-Well Polyvinylidene Fluoride (PVDF) Membranes.

1

Saima Ahmed, Benoit Fatou, Nilesh M. Mehta, Tue B. Bennike, and Hanno Steen

Abstract

Proteomics analysis of urine samples allows for studying the impact of system perturbation. However, meaningful proteomics-based biomarker discovery projects often require the analysis of large patient cohorts with hundreds of samples to describe the biological variability. Thus, robust high-throughput sample processing methods are a prerequisite for clinical proteomics pipelines that minimize experimental bias due to individual sample processing methods. Herein we describe a high-throughput method for parallel 96-well plate-based processing of urine samples for subsequent LC/MS-based proteomic analyses. Protein digestion and subsequent sample processing steps are efficiently performed in 96-well polyvinylidene fluoride (PVDF) membrane plate allowing for the use of vacuum manifolds for rapid liquid transfer, and multichannel pipettes and/or liquid handing robots. In this chapter we make available a detailed step-by-step protocol for our 'MStern blotting' sample processing strategy applied to patient urine samples followed by mass spectrometry-based proteomics analysis. Subsequently, we provide an example application using minimal volume of urine samples (e.g. 150 μL) collected from children pre and post thoracotomy to identify the predominant sites of protein catabolism and aid in the design of therapies to ameliorate protein catabolism and breakdown during critical illness. Furthermore, we demonstrate how the systemic state is reflected in the urine as an easily obtainable, stable, and safe biofluid.

S. Ahmed
Departments of Pathology, Boston Children's Hospital and Harvard Medical School, Boston, MA, USA
e-mail: saima.ahmed@childrens.harvard.edu

B. Fatou · H. Steen (✉)
Departments of Pathology, Boston Children's Hospital and Harvard Medical School, Boston, MA, USA

Precision Vaccines Program, Boston Children's Hospital, Boston, MA, USA
e-mail: benoit.fatou@childrens.harvard.edu; hanno.steen@childrens.harvard.edu

N. M. Mehta
Department of Anesthesiology, Critical Care and Pain Medicine, Boston Children's Hospital, Harvard Medical School, Boston, MA, USA
e-mail: nilesh.mehta@childrens.harvard.edu

T. B. Bennike
Departments of Pathology, Boston Children's Hospital and Harvard Medical School, Boston, MA, USA

Department of Health Science and Technology, Aalborg University, Aalborg, Denmark

Precision Vaccines Program, Boston Children's Hospital, Boston, MA, USA
e-mail: tue.bennike@childrens.harvard.edu

H. M. Baptista Carreira dos Santos (ed.), *Translational Urinomics*, Advances in Experimental Medicine and Biology 1306, https://doi.org/10.1007/978-3-030-63908-2_1

Keywords

Liquid chromatography mass spectrometry · Proteomics · Biomarker · Urine · PVDF · Bottom-up

1.1 Introduction

Mass spectrometry (MS)-based proteomics has increasingly moved into the translational and clinical research arena, where robust and efficient sample processing is prerequisite to process large number of samples. The conventional sample processing methods in proteomics, namely SDS-PAGE- or in-solution-based sample processing, are slow and laborious, and thus do not easily provide the necessary high-throughput. The introduction of filter-aided sample processing method (FASP), initially described by Manza et al. [1] and then fully realized in practice by Wisniewski et al. [2] represented a paradigm shift. These filter-aided methods use ultrafiltration membranes with molecular weight cutoffs (MWCO) in the 10 to 30 kDa range to efficiently remove contaminating small molecules and salts, and to capture denatured proteins. Although the application of FASP in the 96-well plate format has been described [3–5], the major limitation of FASP in the 96-well plate is the much slower speed at which the plates have to be centrifuged, which significantly increases the processing time compared to the individual filters. However, in translational and clinical proteomics, which normally include large cohorts, the 96-well plate is the preferred format for sample processing to minimize the processing-induced variability between the samples.

Here, we provide a detailed step-by-step protocol of a 96-well platecompatible membrane-based proteomic sample processing method, which enables the complete processing of 96 individual urine samples from neat urine to liquid chromatography coupled to MS (LC/MS) ready sample within a single workday. This method uses a hydrophobic polyvinylidene fluoride (PVDF) membrane with 0.45 μm pore. The membrane material ensures efficient capture of the proteins, while the pore size ensures fast liquid transfer through the membrane which in turn significantly reduces sample processing time. Without compromising peptide and protein identification, our method uses a vacuum manifold and circumvents the need for time-consuming centrifugation steps. Additionally, the cost per sample is reduced and the processing method is compatible with standard liquid handling robots. We have previously validated the usage of 'MStern blotting' for projects with hundreds of urine and plasma samples, demonstrating the applicability of the protocol [6–9].

Recent technological advancements with MS acquisition mode has provided superior data quality and quantitation accuracy. Moving forward from traditional data dependent acquisition (DDA) to data independent acquisition (DIA), has provided a more comprehensive proteome coverage including better quantitation and less missing values [8]. It has also been shown that using project specific spectral libraries provides better protein and peptide identification and overall more reliable data [7, 8]. In this chapter we describe yet another biological application of our high throughput method using urine collected from children pre and post-thoracotomy.

The metabolic stress response to surgical insult in humans is characterized by catabolism of the skeletal and visceral protein stores. Free amino acids that are released from the breakdown of proteins are utilized for, (a) protein synthesis to support tissue repair, (b) immune defense, and (c) inflammatory and acute-phase responses Unless optimal protein intake is ensured, children recovering from surgical interventions are often in a net negative protein balance. Both energy and protein adequacy are required, with protein intake of over 1.5 g per kg bodyweight per day to achieve positive protein balance in clinical trials of protein supplementation during pediatric critical illness [10–12]. Persistent negative protein balance results in loss of muscle protein/lean body mass during a vulnerable phase of convalescence. The loss of lean body mass may be associated with significant morbidity and mortality. In adult survivors of critical illness, muscle weakness and disability may persist for up to 5 years after the insult [13]. Preservation of lean body mass has the potential to improve patient outcomes [11]. Hence, optimizing protein delivery in the post-operative phase is an important goal for hospitalized children. Future trials of nutritional and non-nutritive interventions, aimed at preserving the lean body mass, in the post-operative period, are highly desirable. Despite awareness of this metabolic response to stress, the details of its pathophysiology, time course and specific sites of lean body mass depletion have not been adequately described.

The utilization of our high throughput and robust sample processing method was used to investigate the pre and post-thoracotomy urinary proteome. Our findings remarkably found time point specific post-thoracotomy catabolic markers in urine. These findings can elucidate the predominant sites of protein catabolism and aid in the design of therapies to ameliorate protein catabolism and breakdown during critical illness.

1.2 Experimental Section

1.2.1 Materials

Prepare all solutions and buffers using ultrapure water, LC/MS-grade solvents and analytical grade reagents. The following reagents are needed for preparing digestion of a full plate (96 samples) with some solvent to spare for multichannel pipetting. We provide example product numbers for each reagent/consumable to highlight the type and purity used in order to enable the fast and efficient copying of this protocol.

1.2.1.1 Reagents for Digestion

1. Ethanol (Product Number A962-4, Fisher Scientific), 70% in water (Product Number W6500, Fisher Scientific). Mix 14 mL ethanol with 6 mL water.
2. ABC Buffer: 50 mM ammonium bicarbonate (ABC, Product Number A6141-500G, Millipore Sigma). Weigh 0.8 g of ABC and add to a glass beaker. Add water to a volume of 200 mL, and mix. Store at 4 °C to avoid bacterial growth.
3. Urea sample solution: 8 M urea (Product Number U5378-1KG, Millipore Sigma) in ABC buffer. Weigh 48.04 g and add to a glass beaker. Add ABC buffer to a volume of 100 mL and mix (see Note 1).
4. Reducing/DTT stock: 1 M dithiothreitol (DTT, Product Number 43815, Millipore Sigma) in 50 mM ABC. Weigh 0.154 g DTT into a tube in a fume hood and add 50 mM ABC to a volume of 1 mL. Mix, aliquot and store at −20 °C (see Note 2).
5. Reducing DTT solution: 50 mM DTT in urea sample solution. Dilute 0.25 mL DTT stock in 4.75 mL urea sample solution.
6. Alkylating IAA solution: 0.25 M iodoacetamide (IAA, Product Number I1149, Millipore Sigma) in urea sample solution. Weigh 0.23 g IAA into a tube in a fume hood and add urea sample solution to a volume of 5 mL. Mix until all IAA is dissolved (see Note 3).
7. Digestion solution: 5% acetonitrile (ACN, Product Number A955-4, Fisher Scientific), 5% trypsin resuspension buffer (Product Number V5111, Promega) in 50 mM ABC. Mix 4.5 mL 50 mM ABC, 0.25 mL ACN, and 0.25 mL trypsin resuspension buffer (see Note 4).
8. Sequencing grade modified trypsin (Product Number V5111, Promega), 2× 20 μg stock
9. 40% ACN, 0.1% formic acid (FA, Product Number F0507, Millipore Sigma): Mix 23.96 mL water, 16 mL ACN, and 40 μL FA.
10. Loading MS buffer: 5% FA 5% ACN. Mix 0.5 mL FA with 0.5 mL ACN in 9 mL water.

1.2.1.2 Plates and Disposables for Digestion

1. PVDF membrane plate: Multiscreen HTS 0.45 μm Hydrophobic High Protein Binding Membrane 96-well Filtration plate (Product Number MSIPS4510, Millipore-Sigma).
2. 96-well microplate vacuum manifold (Product Number MAVM0960R, Millipore-Sigma).
3. 3× 96-well plates with V-bottom (Product Number 14-222-240, Fisher Scientific), which can accommodate volumes in the 300 μL range.

4. 1× large 96-well plate with V-bottom which can accommodate volumes of >500 μL (Product Number AB-0859, Fisher Scientific) (see Note 5).
5. Disposable reagent reservoirs (Product Number PI15075, Fisher Scientific).
6. Adhesive cover film with holes for the 96-well plates (Product Number 2997-0100, USA Scientific).

1.2.2 Methods (Workflow Design Shown in Fig. 1.1)

Carry out all procedures at room temperature unless otherwise specified, using multichannel pipettes.

1.2.2.1 Prepare Plate with Dry Urea

1. The urine sample must be mixed with dry urea (Product Number U5378, Millipore Sigma). Add urea 1:1 (w/w) to each well, e.g. 150 μg urea for the standard 150 μL urine (see Note 5).

1.2.2.2 Protein Reduction and Alkylation

2. Add 150 μL of neat urine sample (~15 μg of protein) to each well in a prepared plate with urea (see Note 6 and 7).
3. Label a disposable reagent reservoir and add 5 mL reducing DTT solution.
4. Reduce protein disulfide bonds by adding 30 μL reducing DTT solution to each well (final concentration approx. 6.5 mM). Cover the plate with a lid or film to prevent contaminations and/or evaporation.
5. Briefly shake the plate to mix and incubate for 20 min at 27 °C on a Thermomixer set to 1000 rpm.
6. Label a disposable reagent reservoir and add the alkylation solution in a fume hood.
7. Alkylate cysteine residues, by adding 30 μL to each well (final concentration approx. 29 mM).
8. Briefly shake the plate to mix and incubate for 20 minutes at 27 °C on a Thermomixer set at 750 rpm in the dark (e.g. cover with aluminum foil).

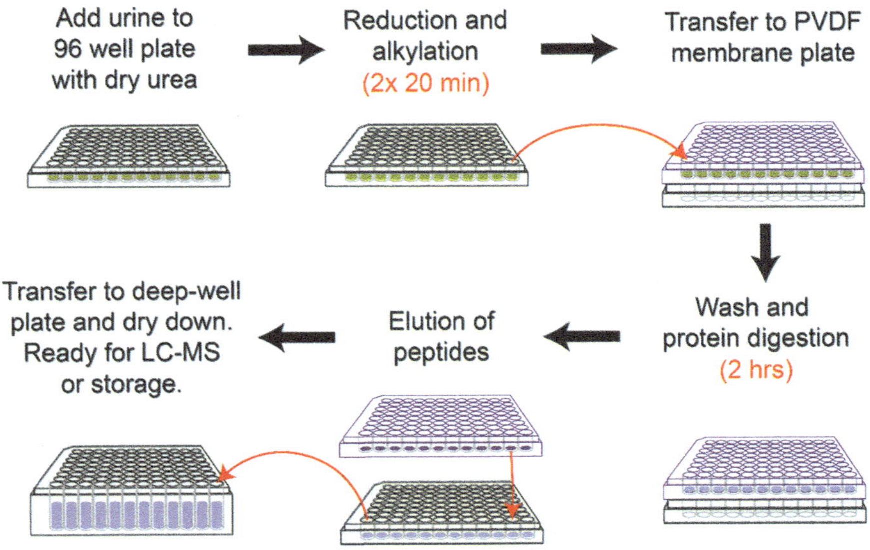

Fig. 1.1 Overview of the high-throughput sample preparation protocol for urine proteomics using 96-well polyvinylidene fluoride (PVDF) membranes plates (depicted in purple)

1.2.2.3 Protein Digestion

1. While the alkylation reaction is ongoing, place the 96-well PVDF membrane plate on a 96-well plate.
2. Label a disposable reagent reservoir and add 20 mL 70% ethanol.
3. Activate the PVDF membrane plate by adding 150 μL 70% ethanol to each well (see Note 8).
4. Place the 96-well PVDF membrane plate with the 96-well collection plate underneath in the vacuum manifold. Slowly apply the vacuum (approx. −1 psi) and allow the liquid to pass through (see Note 8).
5. Discard the flow-through from the 96-well collection plate (see Note 9).
6. Label a disposable reagent reservoir and add 35 mL urea sample solution.
7. Prime the PVDF membrane by adding 300 μL urea sample solution to each well.
8. Pass the urea sample solution through the PVDF membrane using the vacuum manifold and discard the flow-through (see Note 10).
9. Add half of reduced and alkylated urine sample from the 96-well plate to the PVDF membrane 96 well plate and pass the sample using the vacuum.
10. Repeat the step with the flow through, to ensure that all proteins are captured on PVDF membrane and discard the flow through.
11. Add the remaining half of reduced and alkylated urine sample to the PVDF membrane 96 well plate and pass the sample using the vacuum.
12. Repeat the step with the flow through, to ensure that all proteins are captured on PVDF membrane and discard the flow through.
13. Label a disposable reagent reservoir and add 35 mL 50 mM ABC.
14. Wash the filter plate by adding 300 μL 50 mM ABC to each well.
15. Pass the solution through the PVDF membrane using the vacuum manifold and discard the flow-through.
16. Resuspend the dry sequencing grade modified trypsin with 11 mL digestion solution. Label a disposable reagent reservoir and add the digestion solution with trypsin.
17. Place a new collection plate underneath the 96-well PVDF-membrane plate and add 100 μL digestion solution with trypsin to each well.
18. Cover the 96-well PVDF-membrane plate with a plastic/paraffin sheet to prevent evaporation and place the stacked plates in a 37 °C incubator (see Note 11). Incubate the stacked plate for 2 h.
19. Remove the stacked plates from the incubator, place them in the vacuum manifold and transfer the digest into the collection plate by applying vacuum.
20. Add 200 μL of 40% ACN / 0.1% FA to each well in the 96-well PVDF membrane plate and apply the vacuum to elute remaining peptides.
21. Transfer the eluent to a deep 96-well collection plate, to avoid overflowing.
22. Repeat the elution- and transfer step
23. Cover the deep 96-well plate with an adhesive cover film with holes, and evaporate the eluted peptides using a vacuum centrifuge.
24. The dry peptide product can be stored at −20 °C or analyzed immediately.
25. Prior to analysis, resuspend the peptides in an appropriate amount MS loading buffer (e.g. 20 μl), ultra-sonicate the plate for 5 min, mix, and the sample is ready for LC-MS injection.

1.2.3 Data Analysis as Used for the Example Dataset (Monitoring Catabolism Post Thoracotomy by Urine Proteomics)

1.2.3.1 Spectral Library Construction with MaxQuant

1. Create pooled samples by combining a fraction of the sample from similar cohort/groups to limit the number of LC-MS needed for the spectral library. For the example dataset, we created the pools based on the timepoints, each pool containing a small fraction of all samples from that given timepoint.

2. Analyze the samples/pooled samples using LC/MS in DDA mode, with identical LC-parameters as in DIA mode (see Sect. 1.2.3.2 DIA sample Acquisition for LC and MS parameters).
3. Search all DDA RAW files together using MaxQuant with a FASTA file protein database containing all reviewed proteins in the UniProt Human Reference Proteome (downloaded August 21st, 2016, containing 20,210 protein sequence entries) and the sequence for iRT (internal Retention Time) peptides for retention time monitoring.
4. Set MaxQuant parameters to enable LFQ (a label-free protein quantitation analysis was performed using the MaxLFQ algorithm [14], which calculates relative protein abundances based on integrated peptide intensities), match between runs (to allow peptide identifications to be transferred across different LC-MS analyses, based on retention time and accurate m/z). Furthermore, set acetylation of the protein N-termini and oxidation of methionine as variable modifications, and carbamidomethyl of cysteine as a fixed modification. Set maximum number of tryptic missed cleavage to 3. Keep all other parameters as standard including a 1% false discovery rate (FDR) [6, 15, 16].
5. Generate the spectra library in Spectronaut (Biognosys) using default parameters [7]. Optionally, the spectral library can be enlarged using publicly available data which for some projects increases the number of identified proteins and peptides.

Table 1.1 Example DIA method consisting of 375–1200 m/z mass range with 15 variable windows specific to a Q Exactive (Thermo) MS

Window #	Center (m/z)	Window size (m/z)	Min (m/z)	Max (m/z)
0 (MS1)			*375*	*1200*
1	404.75	60.5	374.5	435
2	452.35	36.7	434	470.7
3	485.1	30.8	469.7	500.5
4	513.4	27.8	499.5	527.3
5	540.2	27.8	526.3	554.1
6	568.45	30.7	553.1	583.8
7	597.8	30	582.8	612.8
8	627.55	31.5	611.8	643.3
9	658.8	33	642.3	675.3
10	691.95	35.3	674.3	709.6
11	729.55	41.9	708.6	750.5
12	776.05	53.1	749.5	802.6
13	834.05	64.9	801.6	866.5
14	909.55	88.1	865.5	953.6
15	1075.85	246.5	952.6	1199.1

1.2.3.2 DIA Sample Acquisition

1. Spike each sample with internal retention time (iRT) standard according to iRT kit Quick Reference Card (Product Number Ki-300201 Biognosys).
2. Analyze each sample in DIA mode using the same LC/MS parameters as DDA runs. For solvent in the example dataset, we used 4% ACN in water, 0.1% FA, which was increased to 35% ACN over a 45 min gradient (59 min total run time) at a flowrate of 1 μl/min. We used a 10 cm C18 PicoChip column with a build-in emitter coupled online to a Q Exactive (Thermo Scientific) operating in positive mode.
3. Optimal DIA method set up for urine was in the example dataset found to include: mass range m/z 375–1200, divided into 15 variable sized windows (Table 1.1), 35,000 resolution @ m/z 200. Automatic Gain Control (AGC) target 3e6, maximum Ion Transfer (IT) 130 ms, fixed first mass m/z 200, and Normalized Collision Energy (NCE) 27. The DIA scans are preceded by an MS1 Full scan with identical parameters yielding an actual total cycle time of 2.6 s which is optimal for quantitation of peptides eluting over <15 seconds, which we found was by far the majority at the specified LC parameters [7–9].

1.2.3.3 DIA Data Analysis in Spectronaut

1. Load all DIA raw files into Spectronaut (Biognosys), using default parameters.
2. Analyze the DIA files using the generated spectral library. Spectronaut utilizes the spiked-in HRM peptides for m/z and retention time calibration, and per default filters the results to q-value <0.01 (equal to FDR < 1%). There is no need to setup the sample groupings in Spectronaut if the data analysis is not done in this program.

3. To ensure high quality protein quantitative data, one can consider filtering proteins with less than two quantifiable peptides unique to a protein-group [17]. However, the exact filtering criteria are project specific.
4. Spectronaut can preferably generate Pivot-tables of the resulting protein and peptide data. The identified proteins can subsequently be analyzed in Perseus, R, or other applications, by e.g. importing the Pivot-protein groups (or peptide for a peptide-centric analysis) directly.

1.2.3.4 Data Processing and Statistical Analysis

1. Process all DDA raw files using MaxQuant (the example dataset was analyzed using v1.5.5.1) [14, 18, 19].
2. Process identified proteins in Perseus (the example dataset was analyzed using v1.5.5.3) [20] by first filtering proteins tagged as contaminants and log2 transform all data.
3. Further quantitation can be performed by keeping in at least 70% of the samples in at least one timepoint/replicate/condition.
4. We recommend imputing missing values solemnly for the purpose of conducting principle component analysis. This can be done in Perseus by using numbers drawn from a normal distribution with the standard parameters set in to simulate signals from low abundant proteins.
5. Further, proteins with a statistically significant change of abundance between different conditions can be identified by paired two-samples t-test with permutation-based false positive control, to adjust for multiple hypothesis testing using standard parameters in Perseus (FDR = 0.05, s0 = 0.1).

1.2.4 Notes

1. The urea sample solution should be freshly prepared. The dissolving of urea can take some time. The reaction is endothermic and can be accelerated by gentle heating, e.g. at 20 °C. However, care should be taken not to heat the urea solution above room temperature, as this will accelerate the breakdown of urea resulting in ammonium isocyanate, which in turn leads to carbamylation of proteins and peptides.
2. The reducing stock solution can preferably be aliquoted in 0.5 mL tubes and stored at −20 °C.
3. Iodoacetamide is light sensitive. Keep in the dark when possible. Can be aliquoted and stored at −20 °C.
4. Digestion in the presence of organic solvents (ACN) lowers the abundance of peptides with several missed tryptic cleavage sites.
5. Preparing the 96-well plate with dry urea can preferably be done well in advance. To avoid the need for repeated scale measures, urea sample solution can be added and dried down, prior to addition of the urine sample: Label a disposable reagent reservoir and add 15 mL urea sample solution. Using a multichannel pipette, add 2:1 (v/v) urea sample solution: urine (e.g. 300 μL urea sample solution for 150 μL urine sample) to each well in a 96-well plate using a multichannel pipette. Add an adhesive cover film and dry down in a vacuum centrifuge overnight.
6. To easily keep track of what samples goes into which well, one can preferably download and print a 96-well plate overview [6]. Also, marking the top left corner of the plates can help you keep the orientation.
7. In all transfer-steps, ensure that the A1-well on the 96-well PVDF membrane plate is positioned on the A1-well on the 96-well collection plate. The design might be different. Marking the top left corner of the plates can help you keep the orientation.
8. To avoid damaging the PVDF membranes, care should be taken when applying the vacuum, which should not exceed – 0.5 psi (−1 inHg, −0.034 bar, −3.4 kPa). The liquid should pass through within 10–30 s. Also, the membranes should never be left with high organic solvents for longer than 3 min.
9. Droplets often remain underneath the 96-well PVDF membrane plate following liquid extraction. If this is the case, hold the vacuum manifold with plates inside firmly, and tap them once onto the table. This will cause the droplets to release.

10. It is important that the PVDF membrane does not dry out. Therefore, add the urea samples solution, but do not pass it through before you are ready to load the reduced and alkylated protein sample.
11. We recommend placing a beaker with water in the incubator to ensure sufficient humidity which in turn minimizes evaporation.

1.2.5 Application: Monitoring Catabolism Post Thoracotomy by Urine Proteomics

In this section, we demonstrate a biomedical application of our high-throughput urine sample processing and analysis methods detailed above using paired urine samples collected from 14 children pre thoracotomy (T1), and post thoracotomy (T2 – post-operative sample on the day of surgery, T3 – one day after surgery, T4 – approximately 3 days post-surgery), yielding a total of 56 samples. The objective of this urine proteomics study was to better understand the post-surgery catabolism of the skeletal and visceral protein stores. We also sought to provide novel insight into the urinary proteome changes in children with thoracotomy at different time points. Our hypothesis was that children undergoing major surgery will have significant differences in the proteomic profile before and after surgery. These data could potentially highlight specific types of protein degraded after surgery and provide insights into metabolic pathways that are suppressed or active in the post-operative phase. These observations may guide interventions, such as optimal protein intake, aimed at ameliorating the protein loss after surgery. Using our high-throughput and robust sample processing method coupled with DIA we remarkably found time point specific post-thoracotomy catabolic markers in urine.

In a pilot study, we enrolled children aged 1–18 years who were scheduled for thoracic surgery. Patients with metabolic disorders, diabetes, liver or renal dysfunction, or inability to collect 24-hour urine were excluded. Patients had indwelling urinary catheter as part of routine post-operative management. The study was approved by the institutional review board (IRB P00010517), and written informed consent was obtained from a parent or guardian. 19 patients were enrolled for a stable isotope study of protein turnover, the study protocol and results are described elsewhere [21]. In brief, the cohort with median age 13.2 years underwent thoracotomy or thoracoscopic surgery with an increased protein turnover and net negative balance due to increased protein breakdown. The stable isotope technique demonstrated the whole-body turnover, but the site and metabolic pathways for protein breakdown during the metabolic stress response remained unclear. In 14 patients, pre-operative and post-operative urine samples were obtained for proteomic analysis.

1.3 Results

The 56 urine samples collected from 14 participants at time point T1–4, were processed using the 'MStern blotting' protocol and analyzed by LC/MS in DIA mode. Cumulative, we identified 832 proteins where 623 of them had valid values in at least 70% of the samples in at least one timepoint. Of note: the number of detectable urinary proteins is highly project dependent.

Largest Impact on the Urinary Proteome Day 1 Post Surgery

Since the samples were analyzed in a randomized order, performing an PCA is an excellent analysis tool to visualize any potential batch effects, which becomes more relevant with increasing numbers of samples. Therefore, we performed an unsupervised PCA (Fig. 1.2) to investigate the overall variance in the data. Interestingly, the unsupervised analysis showed 4 groups representing each time point, spite some overlap. Additionally, a trajectory was apparent on the scores plot with samples from later timepoints continuously shift to lower PC-1 values. Our analysis revealed that the main difference in the dataset is described by the timepoint.

We next performed a differential analysis to highlight significant protein changes from the

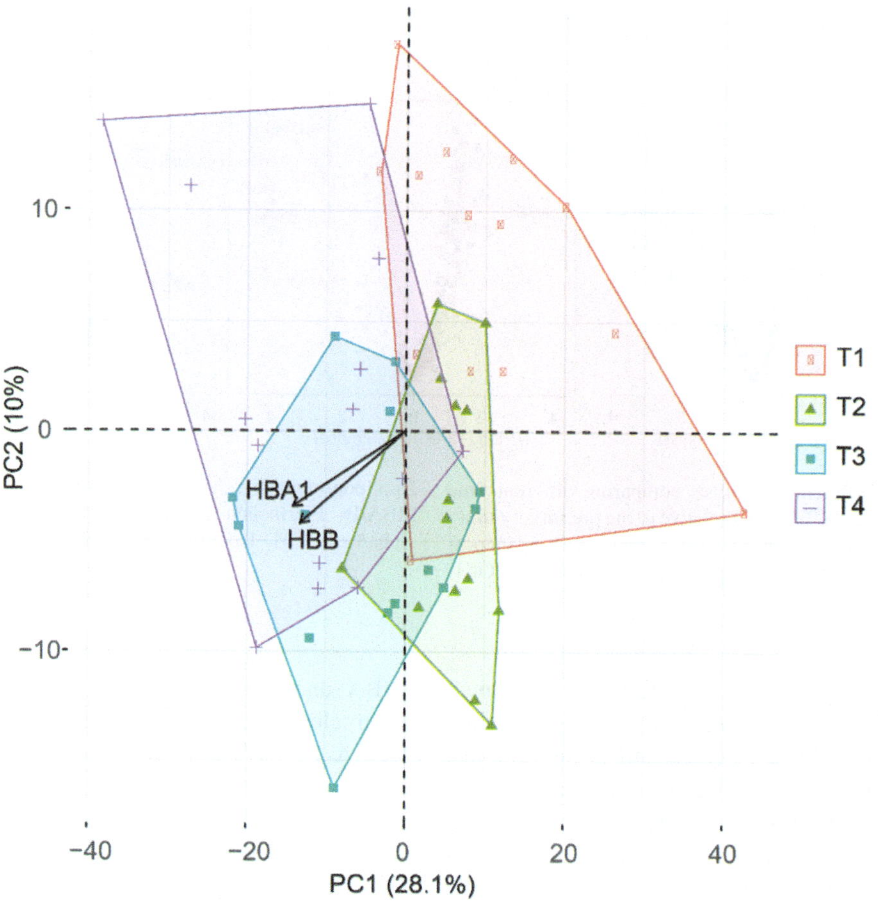

Fig. 1.2 Unsupervised principle component analysis (PCA) scores plot with loadings from hemoglobin alpha (HBA) and beta (HBB), indicating distinct grouping based on timepoint

time points T2, T3, and T4, relative to T1. Comparing the pre-surgery urine (T1) to the urine collected immediately following surgery (T2), no proteins demonstrated a significant abundance change (Fig. 1.3) after performing a paired t-test analysis followed by a false discovery rate-based multiple testing correction. This indicates that the recent surgery had not impacted the urinary proteome composition at the time of collection. In contrast, numerous proteins with significant abundance differences are observed when T3 or T4 is compared to T1. The day after surgery (T3), the hemoglobin subunits as well as some proteins belonging to the complement system (CFB, C4A, CFH, C9) are significantly increased whereas some proteins involved in metabolism (ALDOA, ALDOB, GGT6, PGK1) are significantly decreased compared to the day before the surgery (T1). These changes persist for 3 days after surgery as most of them are still of significantly different abundance when comparing T4 vs. T1.

We noticed a consistent increase in various hemoglobins. However, since there was no evidence of renal injury or dysfunction in any of the patients, we assume that this post-operative increase is due to the trauma from the in-dwelling urinary catheter.

We next performed a protein-protein interaction analysis of the significantly changed proteins (Fig. 1.4). The analysis revealed that clusters of functionally associated proteins were changing in a similar way, e.g. several complement proteins, hemoglobin, aldolases, and S100-proteins.

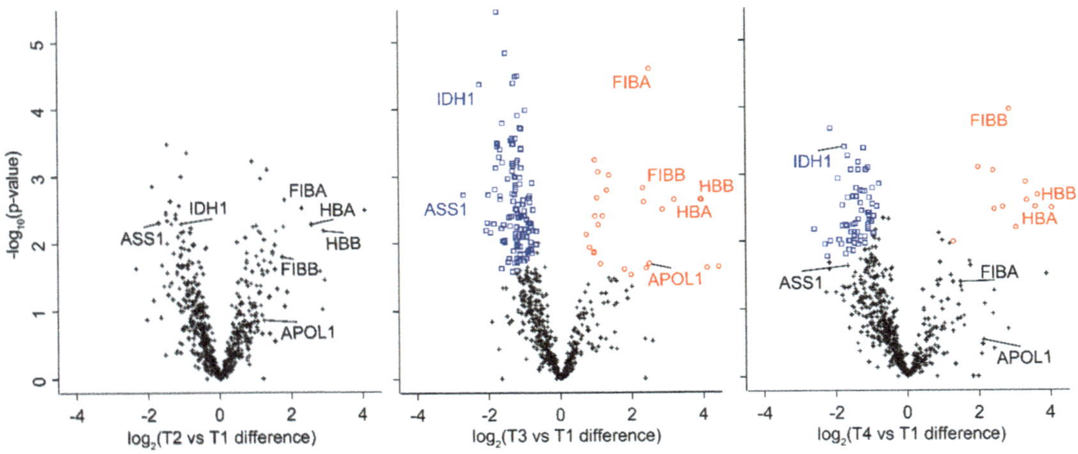

Fig. 1.3 Differential analysis comparing different time points (**a**) T2 (**b**) T3 (**c**) T4 relative to the pre surgery time-point T1. ○: More abundant post T1, □: Less abundant post T1, + not statistically changed (q-value >0.05). APOL1: Apolipoprotein L1, ASS1: Argininosuccinate synthase, FIBA/B: Fibrinogen alpha/beta, HBA/B: Hemoglobin alpha/beta, IDH1 Isocitrate dehydrogenase

Therefore, to identify underlying biological themes, we identified significantly enriched Reactome pathways represented in the proteins. Based on the number of changing proteins, we focused our analysis on the T3-timepoint.

Focusing on the down-regulated proteins, a large majority of them are involved in either the biosynthesis of amino acids, the glycolysis, the carbon metabolism or some other metabolic pathways. All 30 proteins tagged as metabolic pathways (KEGG ID hs01100) were significantly less abundant post T1. The same holds true for the 11 proteins tagged as biosynthesis of amino acids including argininosuccinate synthase (ASS1) (KEGG ID hsa01230) and the 12 proteins tagged as carbon metabolism including isocitrate dehydrogenase (IDH1) (KEGG ID has01200) and glycolysis/gluconeogenesis (KEGG ID hsa00010). Additionally, a number of proteins, although not all, involved in metabolism of proteins (Reactome ID 392499) were increased. This includes apolipoprotein L1 (APOL1) which was increased 5.3× at T3 compared to T1. APOL1 is a part of high-density lipoprotein, which transport lipids in the bloodstream for metabolism. Finally, four proteins, including fibrinogen alpha and beta (FIBA and FIBB), involved in formation of fibrin clot (Clotting Cascade) (Reactome ID R-HSA-140877) were all increased post T1.

Overall our analysis reveals a shift from a carbon metabolism to a fatty acid metabolism post-surgery, including reduced glycolysis, gluconeogenesis, synthesis of amino acids, and increased metabolism of proteins and blood clotting cascades post-surgery. Alterations in substrate utilization, and particularly a preference for using lipoid as a substrate for metabolism, have been previously described in critically ill patients [22, 23] These expected findings validate the applied methodology. The study demonstrates that urine proteomic analysis is feasible and that systemic metabolic changes following stress are reflected well in the urinary proteome. Therefore, urinary proteomic analysis can be studied using the applied methodology. Based on the findings of our pilot study, proteomic analysis may provide important insights into substrate metabolism (preference for fatty acid oxidation) and characteristics of protein breakdown following stress of injury, illness or surgery in children. These observations may guide interventions aimed at ameliorating protein catabolism in this group.

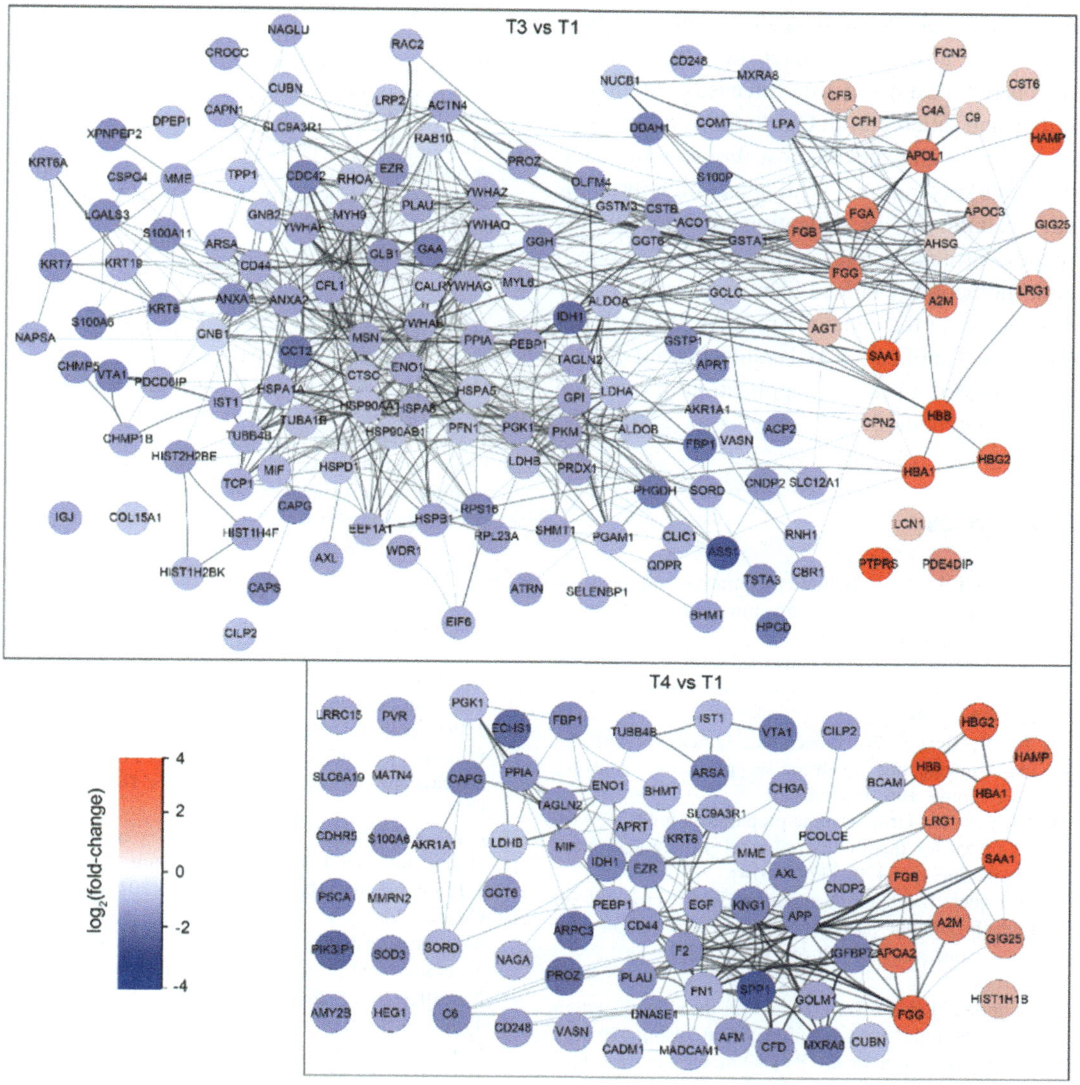

Fig. 1.4 Overview of significantly changed proteins (q-value <0.05), comparing T3, and T4 to T1, respectively. No proteins passed the significance cutoff for T2 compared to T1. Protein-protein interactions from the STRING database, gene-names, and log2-fold change relative to T1 are indicated

1.4 Conclusion

In this chapter we presented a novel and detailed step-by-step protocol of a 96-well plate compatible membrane-based proteomic sample processing method, which enables the complete processing of 96 individual urine samples from the neat urine to the LC/MS ready sample within a single workday. We demonstrated this urine sample processing and analysis using our 'MStern blotting' strategy with a biomedical application. We described an example application of our high-throughput method using minimal volume urine samples (e.g. 150 μL) collected from children pre and post thoracotomy to identify the predominant sites of protein catabolism and aid in the design of therapies to ameliorate protein catabolism and breakdown during critical illness. We also demonstrated how the systemic state is reflected in the urine, an easily obtainable, stable, and safe biofluid. Our findings remarkably found time point specific post-thoracotomy catabolic markers in urine. With newer high throughput technology, identification of time points specific protein species in the urine of children after thoracotomy helped identify unique urinary biomarkers of the catabolic process. Overall, our biomedical application and results thereof clearly demonstrated a practical application of our high throughput method.

References

1. Manza, L.L., et al.: Sample preparation and digestion for proteomic analyses using spin filters. Proteomics. **5**(7), 1742–1745 (2005)
2. Wisniewski, J.R., et al.: Universal sample preparation method for proteome analysis. Nat. Methods. **6**(5), 359–362 (2009)
3. Yu, Y., Bekele, S., Pieper, R.: Quick 96FASP for high throughput quantitative proteome analysis. J. Proteome. **166**, 1–7 (2017)
4. Yu, Y., et al.: Urine sample preparation in 96-well filter plates for quantitative clinical proteomics. Anal. Chem. **86**(11), 5470–5477 (2014)
5. Potriquet, J., et al.: A modified FASP protocol for high-throughput preparation of protein samples for mass spectrometry. PLoS One. **12**(7), e0175967 (2017)
6. Bennike, T.B., Steen, H.: High-throughput parallel proteomic sample preparation using 96-well Polyvinylidene fluoride (PVDF) membranes and C18 purification plates. Methods Mol. Biol. **1619**, 395–402 (2017)
7. Bennike, T.B., et al.: A cost-effective high-throughput plasma and serum proteomics workflow enables mapping of the molecular impact of total pancreatectomy with islet autotransplantation. J. Proteome Res. **17**(5), 1983–1992 (2018)
8. Muntel, J., et al.: Advancing urinary protein biomarker discovery by data-independent acquisition on a quadrupole-orbitrap mass spectrometer. J. Proteome Res. **14**(11), 4752–4762 (2015)
9. Berger, S.T., et al.: MStern blotting-high throughput polyvinylidene fluoride (PVDF) membrane-based proteomic sample preparation for 96-well plates. Mol. Cell. Proteomics. **14**(10), 2814–2823 (2015)
10. Skillman, H.E., Mehta, N.M.: Nutrition therapy in the critically ill child. Curr. Opin. Crit. Care. **18**(2), 192–198 (2012)
11. Mehta, N.M., Compher, C., A.S.P.E.N.B.o. Directors: A.S.P.E.N. Clinical Guidelines: nutrition support of the critically ill child. JPEN J. Parenter. Enteral Nutr. **33**(3), 260–276 (2009)
12. Mehta, N.M., et al.: Severe weight loss and hypermetabolic paroxysmal dysautonomia following hypoxic ischemic brain injury: the role of indirect calorimetry in the intensive care unit. JPEN J. Parenter. Enteral Nutr. **32**(3), 281–284 (2008)
13. Herridge, M.S., et al.: Functional disability 5 years after acute respiratory distress syndrome. N. Engl. J. Med. **364**(14), 1293–1304 (2011)
14. Cox, J., et al.: Accurate proteome-wide label-free quantification by delayed normalization and maximal peptide ratio extraction, termed MaxLFQ. Mol. Cell. Proteomics. **13**(9), 2513–2526 (2014)
15. Bennike, T.B., et al.: Comparing the proteome of snap frozen, RNAlater preserved, and formalin-fixed paraffin-embedded human tissue samples. EuPA Open Proteom. **10**, 9–18 (2016)
16. Bennike, T., et al.: A normative study of the synovial fluid proteome from healthy porcine knee joints. J. Proteome Res. **13**(10), 4377–4387 (2014)
17. Bennike, T.B., et al.: Proteome analysis of rheumatoid arthritis gut mucosa. J. Proteome Res. **16**(1), 346–354 (2017)
18. Cox, J., et al.: Andromeda: a peptide search engine integrated into the MaxQuant environment. J. Proteome Res. **10**(4), 1794–1805 (2011)
19. Cox, J., Mann, M.: MaxQuant enables high peptide identification rates, individualized p.p.b.-range mass accuracies and proteome-wide protein quantification. Nat. Biotechnol. **26**(12), 1367–1372 (2008)
20. Tyanova, S., et al.: The Perseus computational platform for comprehensive analysis of (prote)omics data. Nat. Methods. **13**(9), 731–740 (2016)
21. Fullerton, B.S., et al.: Whole body protein turnover and net protein balance after pediatric thoracic surgery: a noninvasive single-dose (15) N glycine stable isotope protocol with end-product enrichment. JPEN J Parenter Enteral Nutr. **42**(2), 361–370 (2018)
22. Mehta, N.M., Duggan, C.P.: Nutritional deficiencies during critical illness. Pediatr. Clin. N. Am. **56**(5), 1143–1160 (2009)
23. Goode, H.F., et al.: Decreased antioxidant status and increased lipid peroxidation in patients with septic shock and secondary organ dysfunction. Crit. Care Med. **23**(4), 646–651 (1995)

Kidney Diseases: The Age of Molecular Markers

2

Glaucia Luciano da Veiga, Beatriz da Costa Aguiar Alves, Matheus Moreira Perez, Joyce Regina Raimundo, Jéssica Freitas de Araújo Encinas, Neif Murad, and Fernando Luiz Affonso Fonseca

Abstract

Kidney diseases are conditions that increase the morbidity and mortality of those afflicted. Diagnosis of these conditions is based on parameters such as the glomerular filtration rate (GFR), measurement of serum and urinary creatinine levels and equations derived from these measurements (Wasung, Chawla, Madero. Clin Chim Acta 438:350–357, 2015). However, serum creatinine as a marker for measuring renal dysfunction has its limitations since it is altered in several other physiological situations, such as in patients with muscle loss, after intense physical exercise or in people on a high protein diet (Riley, Powers, Welch. Res Q Exerc Sport 52(3):339–347, 1981; Juraschek, Appel, Anderson, Miller. Am J Kidney Dis 61(4):547–554, 2013). Besides the fact that serum creatinine is a marker that indicates glomerular damage, it is necessary the discovery of new biomarkers that reflect not only glomerular damage but also tubular impairment. Recent advances in Molecular Biology have led to the generation or identification of new biomarkers for kidney diseases such as: Acute Kidney Failure (AKI), chronic kidney disease (CKD), nephritis or nephrotic syndrome. There are recent markers that have been used to aid in diagnosis and have been shown to be more sensitive and specific than classical markers, such as neutrophil gelatinase associated lipocalin (NGAL) or kidney injury molecule-1 (KIM-1) (Wasung, Chawla, Madero. Clin Chim Acta 438:350–357, 2015; George, Gounden. Adv Clin Chem 88:91–119, 2019; Han, Bailly, Abichandani, Thadhani, Bonventre. Kidney Int 62(1):237–244, 2002; Fontanilla, Han. Expert Opin Med Diagn 5(2):161–173, 2011). However, early diagnostic biomarkers are still necessary to assist the intervention and monitor of the progression of these conditions.

Keywords

Kidney disease · Biomarkers · Early diagnostic

G. L. da Veiga (✉) · B. da Costa Aguiar Alves
M. M. Perez · J. R. Raimundo
J. F. de Araújo Encinas
Division of Clinical Analysis, Centro Universitário Saúde ABC, Santo André, Brazil

N. Murad
Cardiology Department, Centro Universitário Saúde ABC, Santo André, Brazil

F. L. A. Fonseca
Division of Clinical Analysis, Centro Universitário Saúde ABC, Santo André, Brazil

Pharmaceutical Science Department, Universidade Federal de São Paulo/UNIFESP – Diadema, Butantã, São Paulo, Brazil

H. M. Baptista Carreira dos Santos (ed.), *Translational Urinomics*, Advances in Experimental Medicine and Biology 1306, https://doi.org/10.1007/978-3-030-63908-2_2

2.1 Introduction

2.1.1 Biomarkers

Biomarkers are molecular, biochemical, histological and radiographic characteristics whose measurements are indicative of normal or pathogenic biological processes, in addition to pharmacological processes (response to exposure to treatments or interventions) [7].

The results of biomarker analyses depend not only on the analytical method used, but also on the biology of the biomarker and its cellular origin, the physiology of the associated organ and the pathophysiology of the studied process [8]. Thus, to understand the value of a biomarker, it is necessary to know its pathophysiological relationship with the clinical outcome of a given disease or condition. Good biomarkers should have values and/or measurements whose alterations accompany the variations of the diseases or conditions to which they are associated, thus predicting the clinical outcome of the patient [9].

Biomarkers are classified according to their function in medical practice [9]. *Diagnostic biomarkers* are those that can confirm the presence of a disease or condition, or identify subtypes of this disease, in addition to detecting the presence of pathogens. *Prognostic biomarkers* are capable of tracking disease progression over time to determine the clinical outcome of a patient, while *predictive biomarkers* discern patients who will or will not respond to a specific therapy [7]. Pharmacodynamic biomarkers are subdivided into three different classes: *biomarkers of adverse reactions* that identify patients predisposed to adverse drug reactions even before treatment is initiated [10]; *safety biomarkers* that are capable of identifying early clinical damage caused by the drug [8] or *dose-response biomarkers* that assist in choosing the therapeutic regimen [7].

In several pathological renal situations, such as in suspected renal tumors, diabetic nephropathy or nephritis [11, 12], performing solid biopsy becomes necessary for the diagnosis of the disease. However, it is an invasive procedure and, in many cases, the tissue for analysis is present in sites that are difficult to access. Thus, it is of great clinical importance that less invasive methods are developed and applied throughout the clinical treatment and monitoring process. Recently, liquid biopsies have arousing interest, since they include the analysis of biomarkers for monitoring diseases in biological matrices acquired by a non-invasive (or minimally invasive) technique [13].

2.1.2 Liquid Biopsy

Peripheral blood is the most commonly used biological matrix in different biochemical assessments and in recent years has been used in molecular biology analyses for the diagnosis of chronic diseases, since it has Circulating Tumor Cells (CTCs) and Circulating cell free plasmatic DNA (ccf pDNA) [13]. Cancer patients, for example, have a higher concentration of serum or plasma ccf DNA when compared to healthy people [14].

Urine samples, collected through a totally non-invasive procedure, have proven to be a promising alternative for assessment of renal alterations [15]. Urine is produced in the kidneys and approximately 70% of its proteins and peptides, for example, are derived from this organ [16]. Therefore, the analysis of urinary proteomics apparently reflects the status of the kidneys, so that its use can generate biomarkers for chronic kidney disease (CKD) and other diseases that affect this organ [15]. Even before the concept of liquid biopsy was established, blood and urine had been routinely used for analysis of renal alterations, by analyzing and measuring creatinine, urea, cystatin C and albumin, for example.

As in plasma, urine also contains cell free DNA (ucfDNA), derived from the desquamation of urinary tract cells. UcfDNA is a tool to identify a patient's genomic signature, with great potential for tracking and monitoring urological and non-urological tumors [17], urinary tract infections [18], clinical and experimental nephrotoxicity [19], for example.

Urinary sediment is composed of epithelial cells that have desquamated from the various portions of the urinary tract, from the glomerulus to the urethra. The study of target gene mRNA expression in urinary sediment has been suggested as promising and non-invasive markers of kidney damage [20] and urogenital tumors [21].

Thus, in recent years, several studies have identified and described biomarkers present in serum and urine for acute renal damage and chronic kidney disease [22]. Most of these studies lead to the conclusion that these urinary or serum biomarkers lead to an early diagnosis, with a better prognosis for patients, including a lower kidney transplantation rates or mortality, and thus better quality of life [23]. Next, the particularities of the main kidney diseases and their classical diagnostic biomarkers will be presented, as will markers that are currently being studied.

2.1.3 Nephrotic Syndrome

Nephrotic syndrome is a kidney disease characterized by impairment of glomerular permeability, which prevents the retention of large molecules such as proteins, for example. Thus, large-scale proteinuria appears in these cases, usually above 3.5 g of protein per 1.73 m^2 of body surface in 24-hour urine [24]. This condition may develop in both adults and children and its classification may be primary, in which the cause are different renal alterations, such as:

- Segmental and focal glomerulosclerosis (FSGS): Mesangial cell accumulation is observed and only some glomeruli appear to be injured (focal), as well as only portions of their capillaries (segmental) [25].
- Idiopathic membranous glomerulonephritis (IMGN): predominantly autoimmune in origin, in this type of pathology, IgG4 deposits in the podocytes and increase of this circulating antibody are observed.
- Membranoproliferative glomerulonephritis (MPGN): origin is also predominantly autoimmune. Basal membrane thickening and IgG1 deposits are observed [26].
- Minimal glomerular change (MGC): absence of glomerular lesions or only minimal mesangial lesions, absence of IgM deposits or low level presence. All laboratory tests are altered.
- Glomerulonephritis with Mesangial IgA Deposits (GNIgA): IgA deposits are observed in mesangial cells and less frequently in glomerular capillaries. Glomerular capillary sclerosis and proliferative capillary lesions occur.

Its cause may also be secondary to diseases such as diabetes *mellitus*, lupus, infections, neoplasms and medications [27].

Due to high glomerular permeability, proteins such as antithrombin III and protein C and S are also excreted, which increases the risk of developing venous or arterial thrombosis. Increased albumin excretion triggers hypoalbuminemia and leads to reduced blood oncotic pressure, hypovolemia and edema. The passage of proteins to the glomerular filtrate causes mesangial cell damage, chronic inflammation, increased synthesis and release of cytokines and growth factors, all of which facilitate the deposition of fibrosis and glomerulosclerosis and progressive loss of renal function [24].

2.1.3.1 Diagnosis

The laboratory diagnosis of NS consists of proteinuria above 3.5 g protein per 1.73 m^2 of body surface area in 24 h or above 50 mg/kg of weight in 24 h. The albumin/creatinine ratio is also evaluated, which at values $\geq$3.0 mg/g indicate nephrotic proteinuria. In primary NS, albuminuria values above 3.5 g/24 h or 3.5 g/g creatinine/24 h confirm the presence of nephrotic disease (Veronese et al. 2010), while for NS secondary to diabetic nephropathy (DN), values above 2.2 g/24 and urinary albumin/creatinine ratio of 2.2 mg/g/24 h indicate an excellent correlation for both diagnosis of NS and disease progression [28].

Other important laboratory tests are serum albumin levels, which below 3 g/dL confirms the state of hypoalbuminemia and measurement of total cholesterol and fractions, which usually show increased low-density lipoprotein (LDL). A complete blood count, coagulogram, creatinine, and common urine test are complementary tests that should be evaluated. Kidney ultrasonography is important for assessing tissue size and echogenicity and for signs of acute or chronic kidney failure. The performance of percutaneous renal biopsy is extremely important for histopathological analysis, which will define the etiology of NS, treatment planning and prognosis [24, 29].

2.1.4 Nephritic Syndrome

Nephritic syndrome is defined by the reduction in the glomerular surface available for filtration due to inflammation and/or glomerular injury. It is directly associated with acute clinical manifestations of hematuria, proteinuria, GFR reduction, edema, oliguria (reduction in urinary volume), hypertension and, subsequently, kidney disease [30–32]. It is commonly detected in developing countries because of its association with ineffective or delayed treatments of infectious diseases [33].

This pathology is classified into subtypes, which are as follows: acute poststreptococcal glomerulonephritis (APSGN), caused by Streptococcus and Staphylococcus infections [33], viruses (hepatitis B, mononucleosis and others), protozoa (malaria and toxoplasmosis) and endocarditis; non-streptococcal, related to systemic lupus erythematosus (SLE) [34], IgA nephropathy and other conditions. Acute diffuse glomerulonephritis (ADGN) is in the post streptococcal subclass and is commonly caused by the Class A β-hemolytic Streptococcus (EBHGA) microorganism, generating a renal inflammatory response [33, 35].

Glomerular lesions are mediated by immunological mechanisms classified as: antibody reactions by fixed glomerular antigens or antigens implanted in the glomerulus; and circulating antigen-antibody conjugates (immunocomplexes) that are deposited on the glomerular membrane by the bloodstream. This infiltration of lymphocytes in glomerular regions causes capillary constriction, reduction of the lumen and lesions in the capillary wall that consequently generate an increase in the number of glomerular cells, leading to the proliferation of mesangial and endothelial cells; erythrocyte and protein extravasation that generate hemodynamic changes, electrolyte changes, decreased GFR and accumulation of extracellular fluid (ECF).

2.1.4.1 Diagnosis

Nephritic syndrome is diagnosed through a combination of anamnesis and alterations in classical biochemical markers analyzed in blood and urine samples, such as: macroscopic hematuria or hematic cylinders, erythrocyte dimorphism, proteinuria alterations, blood urea and creatinine, hyperkalemia (high potassium concentrations), acidosis, ANF (diagnosis of SLE), antistreptolysin O (ASLO), anti-DNase (identification of streptococcal bacteria) and cryoglobulins [29].

There is an increase in the number of studies aiming to expand the number of renal biomarkers and predict nephritic syndrome through molecular markers, among which: TGF-β1 (transforming growth factor beta-1) – protein that stimulates cell growth and death, differentiation and accumulation of extracellular matrix in pathological states [36]; BMP-7 (bone morphogenetic protein) - morphogenetic protein highly expressed in medullary tubules, epithelial glomerular cells, podocytes and adventitial cells [11]; and EMT (Epithelial-mesenchymal transition) - highly expressed in this syndrome and responsible for the activation of fibroblasts in kidneys [37].

2.1.5 Acute Kidney Injury

Acute kidney injury (AKI) is described as an acute/abrupt reduction in renal function resulting in: metabolic retention of urea and other nitrogenated compounds, alteration of extracellular volume with fluid accumulation (edema), azotemia (nitrogenous waste), disturbance of acid-base and hydro-electrolytic balance, decreased glomerular filtration and urinary volume [38–40].

AKI progresses rapidly over hours or days with rapid reduction in serum creatinine (SCr) levels and/or glomerular filtration rate; the accelerated and abrupt loss of renal function is what defines the acute character of this pathology [41, 42].

Clinical classification is based on serum creatinine (SCr) levels and urinary volume proposed by the KDIGO group (2012) [38] through stages: Stage 1–1.5 to 2.0 increase in baseline SCr and urinary volume of <0.5 ml/kg/h for 6 h; Stage 2–2.0 to 3.0-fold increase in SCr and urinary volume of <0.5 ml/kg/h >12 h and Stage 3 – >3.0-fold increase in SCr and urinary volume of <0, 3 ml/kg/h for 24 h. The clinical classification of

AKI is commonly categorized as: pre-renal AKI; post-renal AKI; functional and intrarenal AKI.

2.1.5.1 Diagnosis

Conventional diagnostic tests for AKI evaluate: oliguria (abrupt reduction in urinary volume), uremia (presence of urea in the blood), retention of urea, alterations in serum creatinine level, hyperkalemia and metabolic acidosis, reduction in extracellular volume (due to vomiting and diarrhea), anorexia (with retention of urochromes), edema, arterial hypertension and heart failure. Desanti de Oliveira (2019) states that in the last 70 years, the lack of uniform diagnostic and clinical criteria has made the study of the incidence, prevalence and clinical relevance of AKI challenging [41].

2.1.6 Chronic Kidney Disease

Chronic kidney disease (CKD) is considered one of the major public health problems worldwide. In 2016, it affected approximately 47 million adults in the United States (United States Renal Disease Center). It is a progressive and multifactorial pathology characterized by decreased renal function, GFR impairment and renal tissue damage, as well as the presence of parenchymal lesion (Kidney Foundation: Kidney Disease Outcome Quality Initiative [43]).

Clinically, according to the Kidney Foundation: Kidney Disease Outcome Quality Initiative (2002) and the European Best Practice Guidelines Expert Group on Hemodialysis, European Renal Association [44], the patient that presents suboptimal kidney performance for a period longer than three consecutive months is considered to have chronic kidney disease, and this parameter is mainly established by evaluating the glomerular filtration rate (GFR).

CKD can be classified into five stages of worsening severity according to the decline in GFR and the degree of loss of renal function. Renal function index criteria were used to define these stages, and stages 1 and 2 were established with the presence of kidney damage with albuminuria [45], urine sediment, erythrocyte abnormalities and tubular disorders, as well as structural damage. Stages 3 and 4 are defined as having a GFR under 60 ml/min/1.73m^2 with or without a kidney injury marker, or a GFR > 60 ml/min/1.73m^2 associated with a kidney injury marker such as hematuria and/or an albuminuria/proteinuria ratio less than 30 mg/g over a period longer than 3 months. Patients that present a GFR under 15 ml/min/1.73m^2 are in stage 5 of CKD [30, 44, 46, 47]. In the European Union, approximately 13% of the adult population suffers from some degree of CKD, which has a high impact on public spending on treatment [48].

Different factors may be involved in the development of CKD, among which are type II diabetes *mellitus*, chronic glomerulonephritis, and systemic arterial hypertension (SAH) [45, 49]. These factors lead to glomerular hypertension and changes in renal hemodynamics that comprise the most relevant causes for the onset of kidney injury, followed by inflammatory and fibrotic processes and association with oxidative stress [50–52]. Advanced stages of this disease culminate in a process of glomerulosclerosis, glomerular tubule injury, and tubulointerstitial fibrosis, and in the last stage it is necessary to use methods that replace the functions of the kidneys, such as organ transplantation [53–55].

2.1.7 Diabetic Nephropathy

Renal disease stemming from diabetes m*ellitus* is called diabetic nephropathy (DN), which is a glomerular function disorder that, histologically, is due to the presence of glomerulosclerosis accompanied by endothelial dysfunction of afferent and efferent renal arterioles. Generally, it is a silent disease which the physician can only diagnose when there is an effective loss of renal function, at this stage there is no recovery of kidney functionality. Its detection occurs between the moderate and late phases, usually presenting three clinical phases that allow patients to be classified according to disease progression. The first stage is *incipient nephropathy*, characterized by the presence of 30–299 mg of albumin in 24-hour urine (microalbuminuria) which presents an increased cardiovascular risk; the second is *overt nephropathy* with nephrotic proteinuria with over 300 mg of albumin in 24-hour urine (albuminuria) or decreased GFR; and the third is *kidney failure* per se [30, 56, 57].

2.1.7.1 Diagnosis

DN is confirmed by measurement of microalbuminuria (per single sample or sample collection for periods of 4 or 12 night hours), serum creatinine measurement with estimated GFR, in addition to urinalysis including microscopic and urinary sediment analysis. However, these biomarkers are considered inaccurate for being unable to measure actual kidney damage and are relatively insensitive to minor alterations in renal function. However, an early diagnosis is known to guarantee the adoption of preventive measures capable of delaying the evolution of the disease and, consequently, offer the patient a better quality of life [47, 58].

The diagnosis of DN generally follows the same diagnostic protocol as AKI, with the evaluation of renal function being performed using the GFR, measurement of serum creatinine and urea and urinary volume evaluation. The GFR being the best approach for diagnosis. The evaluation of albuminuria also helps to determine renal injury.

New methods are currently being studied that propose the evaluation of biomarkers that can identify minor damage to kidney tissue as well as identify the imminent risk of DN before it is properly diagnosed. Biomarkers such as cystatin C, KIM-1 (Kidney Injury Molecule-1), NGAL (Neutrophil Gelatinase-Associated Lipocalin), Angiotensinogen, Periostin, and MCP-1 (Monocyte Chemoattractant Protein-1) reflect early stage tubular or nephron damage in diabetic patients [59–63].

2.1.8 New Era of Biomarkers and Liquid Matrices

The recent concern with studying less invasive biological matrices for a more accurate and reliable clinical evaluation has led researchers to investigate liquid matrices such as blood, urine, tear fluid and saliva [64–66]. In addition, new collaborative techniques for measuring or detecting biomolecules have improved the accurate diagnosis and prognosis of kidney diseases such as those described above. Thus, we selected the most recent techniques regarding the diagnosis and prognosis of diseases affecting the kidneys. In this scenario, the omic sciences emerge potentially promising in identifying the pathophysiological state of nephropathies [67].

Urine is one of the richest biofluids in metabolites and the preference of studies aiming to identify biomarkers in this matrix is precisely because of the ease of collection, besides the fact that urine reflects the imbalance of the main metabolic pathways in specific pathologies. The metabolic profile of a system, such as the urinary or renal system, is the final step of the omic cascade that is determined by the activation of some genes (genomics) and their transcripts (transcriptome), followed by the production of proteins (proteomics) or metabolites (metabolomics) [68]. Each of these techniques will be briefly outlined below.

Metabolomics is a technique capable of identifying and quantifying metabolic alterations in different tissues, cells or biological fluids. This technique makes it possible to measure small molecules (<1500 daltons), such as sugars, amino acids, fatty acids, nucleotides and lipids in a cell or tissue. At this evaluation level, alterations in the metabolic profile of a system are identified even before there are clinical symptoms. The evaluation of this profile is performed by high throughput nuclear magnetic resonance spectroscopy (NMR) or mass spectroscopy (MS), with these methods it is possible to draw a complete profile of the metabolites in a sample through an analytical spectrum and to draw comparisons between them using specific statistical techniques that recognize patterns. These patterns are like fingerprints and combined they will be compared with others, facilitating the understanding of the metabolic pathway that is altered and the actual cell conditions at the time of collection [69]. A human metabolite database (HMDB) includes all metabolites so far identified, their biofunctions, metabolic pathways, and normal or abnormal concentrations [70]. Nevertheless, researchers in this field are still cautious to use it in medical routines because, unlike the study of genes and proteins, metabolites are not specific to a single pathway [71]. Assessing the metabolic profile in kidney diseases is complex. In the clinical context, decreased renal function affects circulating metabolite levels. The influence of the kidneys on these levels is due to some mechanisms such as glomerular filtration, tubular secretion and catabolism, among others [72]. In a study analyzing the urine of 158 diabetic volunteers with and without CKD, a variety of biomarkers were observed in the samples, such as cytochrome C oxidase and PPARγ-coactivator-1alpha (PGC1α) mRNA, for example. Among the metabolites present in urine it is also possible to quantify mtDNA in exosomes to measure mitochondrial activity in patients with kidney disease [73].

Proteomics is a qualitative and quantitative study of the proteome performed by mass spectrometry. This technique allows the description of cell and tissue differentiation. The identification of protein structures can be done in 3D by means of X-ray, NMR and cryo-electron microscopy. Data obtained through the study of proteomics are deposited in protein lists and metadata in public databases such as PRoteomics IDEntifications [74]. In this way, the genesis of proteomic evaluation made possible the global study of protein expression. Siwy et al. (2017) used the capillary electrophoresis coupled to mass spectrometry (CE-MS) technique which analyzes specific proteins (biomarkers) capable of discriminating the different phases of CKD or other renal alterations. As a result, the authors identified 619 renal disease-specific biomarkers in urine samples [75]. These markers can be used as adjuvants or determinants in the diagnosis of diseases such as AKI, for example, and this can directly reflect on the patient's survival.

Genomics evaluates the whole genome of the individual. The human genome has approximately 3 billion base pairs of DNA and is capable of coding over 20,000 different genes. Within this potential for vast amounts of information it is known that only 1–2% of the entire genome is part of coding regions, with the remaining 98% making up non-coding regions [76, 77]. Considering this wide range of information, it is easy to understand how this science is highlighted in the study of genetic factors that determine the presence of some diseases in the individual, including kidney diseases. This methodology allows the extraction of diagnostic and prognostic information in addition to directing preventive and therapeutic measures for numerous diseases. Some branches of this science include computational genomics, functional genomics, and post-genomic sciences that evaluate the expression of genes that determine the synthesis of specific proteins [78]. The genomic study is able to identify pathological (mutations) and non-pathological (polymorphisms) genetic variations [76]. Inherited etiologies are known to account for approximately 10% of end-stage kidney disease in adults and over 70% of cases of pediatric nephropathy. Therefore, genetic diagnosis is capable of providing accurate information about the pathogenesis of nephrological disorders and can be a determining factor in the diagnosis and clinical decision support. Genetic testing in patients with kidney disease is also used to identify the inherited form of the condition and to assess the compatibility of living kidney donors [79]. Another possible evaluation is the quantification of DNA methylation, which is a widespread epigenetic study in recent years [80]. DNA extraction from fresh or frozen urine samples from patients with kidney disease makes the study of DNA methylation possible, but the success rate of extraction of this material in urine is still modest and needs to be solidified in order to perform this technique in the medical routine [79].

Finally, accurate studies of gene expression of molecules involved in the onset of kidney disease are on the rise and have shown greater accuracy compared to most classical biochemical analyses in diagnostic practice. Sadar et al. (2016) evaluated the expression of KIM-1, NGAL, TGF-β and type I collagen genes in streptozotocin-induced diabetic rats. In this study the authors treated the animals with L-glutamine, an amino acid with antidiabetic and antioxidant potential, and found that the rats showed improvement in biochemical parameters related to the onset of kidney disease. Expression of the studied genes in the kidneys was inhibited after treatment, confirming the effectiveness of the study of mRNA of these genes [81]. Kaucsár et al. (2016) studied the accuracy of NGAL gene expression as a marker of acute kidney failure in mice submitted to different degrees of renal ischemia. A 10-minute ischemia resulted in a high expression of the NGAL gene in the kidney and protein expression of NGAL in the blood and urine increased in parallel with gene expression in kidney tissue, confirming the high sensitivity and specificity of this marker in the identification of subclinical AKI [82]. A clinical study evaluated the reliability of predicting the transition of AKI to CKD, and investigated the potential prognosis of urinary biomarkers such as NGAL (protein expression) and biomarkers KIM-1, IL-18, alpha-1-microglobulin (α1M), sodium/hydrogen exchanger-3 (NHE-3), beta-2 microglobulin (β2M) and acetyl-β-D-glucosaminidase (NAG) by gene expression in urinary sediment. In this study urinary NGAL levels correlated with serum creatinine. β2M expression was inversely proportional to GFR, while α1M and NHE-3 expressions were increased in cases of ischemia-mediated tubular necrosis, demonstrating that the evaluation of these markers in patients transitioning from AKI to CKD is accurate. Therefore, quantification of urinary mRNA can be used as a non-invasive tool in risk stratification for these pathologies [83].

Considering these modern and effective techniques for measuring and diagnosing kidney diseases, new biomarkers are being proposed to assist in the early detection of these diseases applied to these techniques. These are molecules capable of detecting the onset of pathology in time to prevent the disease from progressing and thus improve the quality of life and survival of the nephropathic patient. For this, the biomarker must be highly sensitive and specific, thus providing good accuracy. Figure 2.1 and Table 2.1 shows the most recent biomarkers of kidney disease studied in last few years.

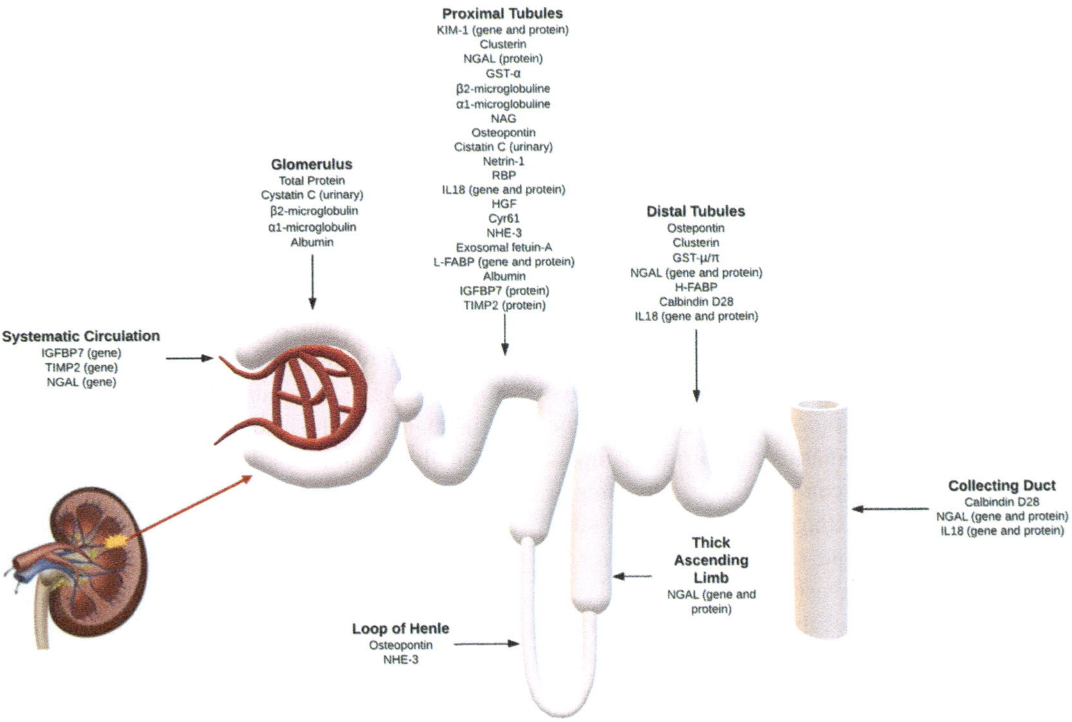

Fig. 2.1 Representative scheme of nephron segments and lesion biomarkers

Table 2.1 List of major kidney disease biomarkers studied in the last decade

New Biomarkers in kidney diseases [84–134]				
Biomarker	Function/kidney	Nephron Segment	Sample	Diseases
Calbindin D28	Vitamin D-dependent clacium-binding protein – reabsorption of calcium and magnesium	Principal cells of distal tubule and Cortical duct collecting	Urine	Drug-induced kidney injury
Clusterin or apolipoprotein J (chaperone)	Tissue remodeling, membrane recycling, stabilization of stressed proteins, and is an inhibitor of apoptosis	Proximal tubule Distal tubule	Urine Serum	Kidney damage (nonspecific) Ischemic AKI
COL IV	Extracellular matrix component	Glomerulus and renal interstitium	Urine Serum	Glomerular diseases Chronic kidney disease (initial and end stage) Diabetic nephropathy
Cysteine-rich protein 61 (Cyr61)	Cysteine-rich matricellular protein Extracellular matrix formation Cell adhesion Proliferation, differentiation Angiogenesis, Apoptosis and inflammation	Proximal tubule Thick ascending limb of Henle's loop Distal and colleting tubule Podocytes	Urine	AKI Membranous nephropathy Diabetic nephropathy IgA nephropathy
Fetuin A	Glycoprotein (secreted from adipose and hepatic tissues) Calcification of vessels	Proximal tubule	Urine Serum	Chronic kidney disease Glomerulopathy AKI Diabetes nephropathy Nephrolithiasis
π-Glutathione S-Transferase (GST- π)	Enzymes of a multigene family	Distal tubule damage	Urine	AKI
Heart type of fatty acid-binding protein (H-FABP)	Cytoplasmic protein	Proximal Distal tubule	Urine Serum	AKI Nephrotoxic AKI Chronic kidney disease
Hepatocyte growth factor (HGF)	Pleiotropic cytokine Angiogenic and angioprotective actions Tubulogenesis	Mesangial cells Endothelial cells Proximal tubule Distal tubule	Urine Serum	AKI Chronic kidney disease Nephrolithiasis
Insulin-like growth factor-binding protein 7 (IGFBP-7)	Cell cycle arrest protein	Renal tubular epithelial cells	Urine Serum	AKI Chronic kidney disease
Interleukin-18 (IL-18)	Cytokine	Proximal tubule	Urine Serum	AKI Chronic kidney disease
Kidney injury Molecule-1 (KIM-1)	Type 1 cell membrane glycoprotein	Proximal tubule	Urine Plasma	AKI Chronic kidney disease Diabete nephropathy End stage of kidney disease

(continued)

Table 2.1 (continued)

New Biomarkers in kidney diseases [84–134]				
Biomarker	Function/kidney	Nephron Segment	Sample	Diseases
Liver-type of fatty acid-binding protein (L-FABP)	Cytoplasmic protein	Lysosomal compartment proximal tubule	Urine Serum	Ischemic and nephrotoxin-induced AKI Chronic kidney disease Diabetic nephropathy Glomerular nephritis
N-acetyl-β-D glucosaminidase (NAG)	Enzyme found predominantly in lysosomes of the proximal tubular cells	Proximal tubule	Urine Blood	Diabetic nephropathy Primary glomerulonephritis Idiopathic membranous nephropathy Primary focal segmental glomerulosclerosis Sepsis (AKI)
Netrin-1	Conserved family of laminin-related Proteins Angiogenesis Cell migration, tissue morphogenesis Regulation of inflammation	Proximal tubule	Urine Serum	Chronic kidney disease AKI
Neutrophil gelatinase-associated lipocalin (NGAL)	Membrane glycoprotein Bacteriostatic agente	Proximal and distal tubules Ascending limb of the loop of Henle	Urine Serum	Chronic kidney disease Acute kidney injury Autossomal polycystic kidney disease Glomerulonephritis Autosomal dominant polycystic kidney disease
Na^+/H^+ exchanger isoform 3 (NHE3)	Apical sodium transporter	Apical membrane and subapical endosomes of renal proximal tubular cells Apical membrane of thick ascending limb cells	Urine	AKI Prerenal azotemia
Osteopontin	Pleiotropic glycoprotein	Thick ascending limbs of the loop of Henle Distal tubules	Urine Serum	Chronic kidney disease Diabetes nephropathy AKI
Retinol binding protein (RBP)	Vitamin A transport	Proximal tubule	Urine Serum	AKI
Tissue inhibitor of MetalloProteinases-2 (TIMP-2)	Cell cycle arrest	All epithelial cells	Urine	AKI Chronic kidney disease
α1-antitrypsin	Member of the serpin superfamily Potent inhibitor of multiple serine proteinases	Tubule epitelial cells	Urine Serum	AKI Chronic kidney disease
α1-microglobulin	Member of lipocalin family	Proximal tubule dysfunction	Urine	Fabry nephropathy Nephrotoxic AKI Renal transplantation Acute tubulointerstitial nephritis Chronic kidney disease
β2-microglobulin (β_2M)	Light chain component of the major histocompatibility class I	Proximal tubule	Urine Serum	Rhabdomyolysis Drug nephrotoxicity AKI

References

1. Wasung, M.E., Chawla, L.S., Madero, M.: Biomarkers of renal function, which and when? Clin. Chim. Acta. **438**, 350–357 (2015)
2. Riley, W.W., Powers, S.K., Welch, H.G.: The effect of two levels of muscular work on urinary creatinine excretion. Res. Q. Exerc. Sport. **52**(3), 339–347 (1981)
3. Juraschek, S.P., Appel, L.J., Anderson, C.A., Miller, E.R.: Effect of a high-protein diet on kidney function in healthy adults: results from the OmniHeart trial. Am. J. Kidney Dis. **61**(4), 547–554 (2013)
4. George, J.A., Gounden, V.: Novel glomerular filtration markers. Adv. Clin. Chem. **88**, 91–119 (2019)
5. Han, W.K., Bailly, V., Abichandani, R., Thadhani, R., Bonventre, J.V.: Kidney Injury Molecule-1 (KIM-1): a novel biomarker for human renal proximal tubule injury. Kidney Int. **62**(1), 237–244 (2002)
6. Fontanilla, J., Han, W.K.: Kidney injury molecule-1 as an early detection tool for acute kidney injury and other kidney diseases. Expert Opin Med Diagn. **5**(2), 161–173 (2011)
7. Califf, R.M.: Biomarker definitions and their applications. Exp. Biol. Med. (Maywood). **243**(3), 213–221 (2018)
8. Sasseville, V.G., Mansfield, K.G., Brees, D.J.: Safety biomarkers in preclinical development: translational potential. Vet. Pathol. **51**(1), 281–291 (2014)
9. Aronson, J.K., Ferner, R.E.: Biomarkers – a general review. Curr Protoc Pharmacol. **76**, 9.23.1–9.23.17 (2017)
10. Carr, D.F., Pirmohamed, M.: Biomarkers of adverse drug reactions. Exp. Biol. Med. (Maywood). **243**(3), 291–299 (2018)
11. Patel, S.R., Dressler, G.R.: BMP7 signaling in renal development and disease. Trends Mol. Med. **11**(11), 512–518 (2005)
12. Zhang, L., Li, X.S., Zhou, L.Q.: Renal tumor biopsy technique. Chin. Med. J. **129**(10), 1236–1240 (2016)
13. Pantel, K., Alix-Panabières, C.: Liquid biopsy and minimal residual disease – latest advances and implications for cure. Nat. Rev. Clin. Oncol. **16**(7), 409–424 (2019)
14. Delgado, P.O., Alves, B.C., Gehrke, F.S., Kuniyoshi, R.K., Wroclavski, M.L., Del Giglio, A., et al.: Characterization of cell-free circulating DNA in plasma in patients with prostate cancer. Tumour Biol. **34**(2), 983–986 (2013)
15. Jain, S., Lin, S.Y., Song, W., Su, Y.H.: Urine-based liquid biopsy for nonurological cancers. Genet. Test. Mol. Biomarkers. **23**(4), 277–283 (2019)
16. Thongboonkerd, V., Malasit, P.: Renal and urinary proteomics: current applications and challenges. Proteomics. **5**(4), 1033–1042 (2005)
17. Salvi, S., Casadio, V.: Urinary cell-free DNA: potential and applications. Methods Mol. Biol. **1909**, 201–209 (2019)
18. Burnham, P., Dadhania, D., Heyang, M., Chen, F., Westblade, L.F., Suthanthiran, M., et al.: Urinary cell-free DNA is a versatile analyte for monitoring infections of the urinary tract. Nat. Commun. **9**(1), 2412 (2018)
19. Le Lann, A.D., Bret, L.C., Delverdier, M.E., Amardeilh, M.F., Alary, C., Fournié, G.J.: Urinary DNA as an indicator of nephrotoxicity caused by endotoxin and gentamicin in mice. Toxicology. **92**(1–3), 1–10 (1994)
20. Szeto, C.C., Chow, K.M., Lai, K.B., Szeto, C.Y., Chan, R.W., Kwan, B.C., et al.: mRNA expression of target genes in the urinary sediment as a noninvasive prognostic indicator of CKD. Am. J. Kidney Dis. **47**(4), 578–586 (2006)
21. Mengual, L., Lozano, J.J., Ingelmo-Torres, M., Izquierdo, L., Musquera, M., Ribal, M.J., et al.: Using gene expression from urine sediment to diagnose prostate cancer: development of a new multiplex mRNA urine test and validation of current biomarkers. BMC Cancer. **16**, 76 (2016)
22. Coca, S.G., Yalavarthy, R., Concato, J., Parikh, C.R.: Biomarkers for the diagnosis and risk stratification of acute kidney injury: a systematic review. Kidney Int. **73**(9), 1008–1016 (2008)
23. Vanmassenhove, J., Glorieux, G., Hoste, E., Dhondt, A., Vanholder, R., Van Biesen, W.: Urinary output and fractional excretion of sodium and urea as indicators of transient versus intrinsic acute kidney injury during early sepsis. Crit. Care. **17**(5), R234 (2013)
24. Khanna, R.: Clinical presentation & management of glomerular diseases: hematuria, nephritic & nephrotic syndrome. Mo Med. **108**(1), 33–36 (2011)
25. Fogo, A.B.: Causes and pathogenesis of focal segmental glomerulosclerosis. Nat. Rev. Nephrol. **11**(2), 76–87 (2015)
26. Shimohata, H., Ohgi, K., Maruyama, H., Miyamoto, Y., Takayashu, M., Hirayama, K., et al.: A case of proliferative glomerulonephritis with monoclonal IgG deposits that showed predominantly membranous features. Case Rep Nephrol. **2017**, 1027376 (2017)
27. Orr, P., Shank, B.C., Hickson, S., Cooke, J.: Clinical management of glomerular diseases. Nurs. Clin. North Am. **53**(4), 551–567 (2018)
28. Stoycheff, N., Stevens, L.A., Schmid, C.H., Tighiouart, H., Lewis, J., Atkins, R.C., et al.: Nephrotic syndrome in diabetic kidney disease: an evaluation and update of the definition. Am. J. Kidney Dis. **54**(5), 840–849 (2009)
29. Madaio, M.P., Harrington, J.T.: The diagnosis of glomerular diseases: acute glomerulonephritis and the nephrotic syndrome. Arch. Intern. Med. **161**(1), 25–34 (2001)
30. Levey, A.S., de Jong, P.E., Coresh, J., El Nahas, M., Astor, B.C., Matsushita, K., et al.: The definition, classification, and prognosis of chronic kidney disease: a KDIGO controversies conference report.

Kidney Int. **80**(1), 17–28 (2011 July). https://doi.org/10.1038/ki.2010.483. Epub 2010. Dec 8
31. Mason, P.D., Pusey, C.D.: Glomerulonephritis: diagnosis and treatment. BMJ. **309**(6968), 1557–1563 (1994)
32. Beck, L.H., Salant, D.J.: Glomerular and tubulointerstitial diseases. Prim. Care. **35**(2), 265–296, vi (2008)
33. Ferretti, J.J., Stevens, D.L., Fischetti, V.A.: Basic Biology to Clinical Manifestations. 2016
34. Yu, F., Haas, M., Glassock, R., Zhao, M.H.: Redefining lupus nephritis: clinical implications of pathophysiologic subtypes. Nat. Rev. Nephrol. **13**(8), 483–495 (2017)
35. Becker, C.G., Murphy, G.E.: The experimental induction of glomerulonephritis like that in man by infection with group A streptococci. J. Exp. Med. **127**(1), 1–24 (1968)
36. Kanitakis, J., Thivolet, J.: Cyclosporine. An immunosuppressant affecting epithelial cell proliferation. Arch. Dermatol. **126**(3), 369–375 (1990)
37. Lan, H.Y.: Tubular epithelial-myofibroblast transdifferentiation mechanisms in proximal tubule cells. Curr. Opin. Nephrol. Hypertens. **12**(1), 25–29 (2003)
38. Kirsztajn, G.M., Filho, N.S., Draibe, S.A., Netto, M.V., Thomé, F.S., Souza, E., et al.: Fast reading of the KDIGO 2012: guidelines for evaluation and management of chronic kidney disease in clinical practice. J. Bras. Nefrol. **36**(1), 63–73 (2014)
39. Ronco, C., Rizo-Topete, L., Serrano-Soto, M., Kashani, K.: Pro: prevention of acute kidney injury: time for teamwork and new biomarkers. Nephrol. Dial. Transplant. **32**(3), 408–413 (2017)
40. Ronco, C., Bellomo, R., Kellum, J.A.: Acute kidney injury. Lancet. **394**(10212), 1949–1964 (2019)
41. Desanti De Oliveira, B., Xu, K., Shen, T.H., Callahan, M., Kiryluk, K., D'Agati, V.D., et al.: Molecular nephrology: types of acute tubular injury. Nat. Rev. Nephrol. **15**(10), 599–612 (2019)
42. Rangaswamy, D., Sud, K.: Acute kidney injury and disease: long-term consequences and management. Nephrology (Carlton). **23**(11), 969–980 (2018)
43. K/DOQI clinical practice guidelines for chronic kidney disease: evaluation, classification, and stratification. Am J Kidney Dis. 2002;39(2 Suppl 1):S1–266
44. Section, I.: Measurement of renal function, when to refer and when to start dialysis. Nephrol. Dial. Transplant. **7**, 7–15 (2002)
45. Klahr, S., Morrissey, J.: Progression of chronic renal disease. Am. J. Kidney Dis. **41**(3 Suppl 1), S3–S7 (2003)
46. Levey, A.S., Coresh, J.: Chronic kidney disease. Lancet. **379**(9811), 165–180 (2012)
47. Miller, W.G., Bachmann, L.M., Delanghe, J.R., Inker, L.A., Jones, G.R.D., Vassalotti, J.A.: Optimal use of biomarkers for chronic kidney disease. Clin. Chem. **65**(8), 949–955 (2019)
48. Kent, S., Schlackow, I., Lozano-Kuhne, J., Reith, C., Emberson, J., Haynes, R., et al.: What is the impact of chronic kidney disease stage and cardiovascular disease on the annual cost of hospital care in moderate-to-severe kidney disease? BMC Nephrol. **16**(65), 015–0054 (2015)
49. Woo, K.T., Wong, K.S., Chan, C.M.: Clinical trials of the past decade in the management of chronic kidney disease. Rev. Recent Clin. Trials. **4**(3), 159–162 (2009)
50. Kujal, P., Vernerova, Z.: 5/6 nephrectomy as an experimental model of chronic renal failure and adaptation to reduced nephron number. Cesk. Fysiol. **57**(4), 104–109 (2008)
51. Taal, M.W., Zandi-Nejad, K., Weening, B., Shahsafaei, A., Kato, S., Lee, K.W., et al.: Proinflammatory gene expression and macrophage recruitment in the rat remnant kidney. Kidney Int. **58**(4), 1664–1676 (2000)
52. Sasaki, M., Shikata, K., Okada, S., Miyamoto, S., Nishishita, S., Kataoka, H.U., et al.: The macrophage is a key factor in renal injuries caused by glomerular hyperfiltration. Acta Med Okayama. **65**(2), 81–89 (2011)
53. Remuzzi, A., Gagliardini, E., Sangalli, F., Bonomelli, M., Piccinelli, M., Benigni, A., et al.: ACE inhibition reduces glomerulosclerosis and regenerates glomerular tissue in a model of progressive renal disease. Kidney Int. **69**(7), 1124–1130 (2006)
54. Yoshida, Y., Fogo, A., Glick, A.D., Ichikawa, I.: Biology of progression of chronic renal diseases. Keio J. Med. **36**(4), 369–380 (1987)
55. Waldherr, R., Gretz, N.: Natural course of the development of histological lesions after 5/6 nephrectomy. Contrib. Nephrol. **60**, 64–72 (1988)
56. Romero, C., Caballero, N., Herrero, M., Ruíz, R., Sadick, N.S., Trelles, M.A.: Effects of cellulite treatment with RF, IR light, mechanical massage and suction treating one buttock with the contralateral as a control. J. Cosmet. Laser Ther. **10**(4), 193–201 (2008)
57. Romero-Aroca, P.: Targeting the pathophysiology of diabetic macular edema. Diabetes Care. **33**(11), 2484–2485 (2010)
58. Norris, K.C., Smoyer, K.E., Rolland, C., Van der Vaart, J., Grubb, E.B.: Albuminuria, serum creatinine, and estimated glomerular filtration rate as predictors of cardio-renal outcomes in patients with type 2 diabetes mellitus and kidney disease: a systematic literature review. BMC Nephrol. **19**(1), 36 (2018)
59. Satirapoj, B.: Tubulointerstitial biomarkers for diabetic nephropathy. J. Diabetes Res. **2018**, 2852398 (2018)
60. Karmin, O., Siow, Y.L.: Metabolic imbalance of homocysteine and hydrogen sulfide in kidney disease. Curr. Med. Chem. **25**(3), 367–377 (2018)
61. Villeval'de, S.V., Gudgalis, N.I., Kobalava, Z.D.: Cistatin C as a novel marker of renal function impairement and cardiovascular risk. Kardiologiia. **50**(6), 78–82 (2010)
62. Żyłka, A., Dumnicka, P., Kuśnierz-Cabala, B., Gala-Błądzińska, A., Ceranowicz, P., Kucharz, J.,

et al.: Markers of glomerular and tubular damage in the early stage of kidney disease in type 2 diabetic patients. Mediat. Inflamm. **2018**, 7659243 (2018)

63. Kapoula, G.V., Kontou, P.I., Bagos, P.G.: Diagnostic accuracy of neutrophil gelatinase-associated Lipocalin for predicting early diabetic nephropathy in patients with type 1 and type 2 diabetes mellitus: a systematic review and meta-analysis. J Appl Lab Med. **4**(1), 78–94 (2019)
64. Mayer, G., Heerspink, H.J., Aschauer, C., Heinzel, A., Heinze, G., Kainz, A., et al.: Systems biology-derived biomarkers to predict progression of renal function decline in type 2 diabetes. Diabetes Care. **40**(3), 391–397 (2017)
65. Lai, Y.T., Cerquinho, R.G., Perez, M.M., Alves, B.D.C.A., Pereira, E.C., Azzalis, L.A., et al.: Determination of vitamin D in tears of healthy individuals by the electrochemiluminescence method. J. Clin. Lab. Anal. **33**(4), e22830 (2019)
66. Luz, M.C., Perez, M.M., Azzalis, L.A., Sousa, L.V., Adami, F., Fonseca, F.L., et al.: Evaluation of MCT1, MCT4 and CD147 genes in peripheral blood cells of breast cancer patients and their potential use as diagnostic and prognostic markers. Int. J. Mol. Sci. **18**(4) (2017)
67. Manzoni, C., Kia, D.A., Vandrovcova, J., Hardy, J., Wood, N.W., Lewis, P.A., et al.: Genome, transcriptome and proteome: the rise of omics data and their integration in biomedical sciences. Brief. Bioinform. **19**(2), 286–302 (2018)
68. Khamis, M.M., Adamko, D.J., El-Aneed, A.: Mass spectrometric based approaches in urine metabolomics and biomarker discovery. Mass Spectrom. Rev. **36**(2), 115–134 (2017)
69. Schmidt, C.: Metabolomics takes its place as latest up-and-coming "omic" science. J. Natl. Cancer Inst. **96**(10), 732–734 (2004)
70. Wishart, D.S., Jewison, T., Guo, A.C., Wilson, M., Knox, C., Liu, Y., et al.: HMDB 3.0--the human metabolome database in 2013. Nucleic Acids Res. **41**(Database issue), D801–D807 (2013)
71. Kumar, B.S., Chung, B.C., Kwon, O.S., Jung, B.H.: Discovery of common urinary biomarkers for hepatotoxicity induced by carbon tetrachloride, acetaminophen and methotrexate by mass spectrometry-based metabolomics. J. Appl. Toxicol. **32**(7), 505–520 (2012)
72. Rhee, E.P., Clish, C.B., Ghorbani, A., Larson, M.G., Elmariah, S., McCabe, E., et al.: A combined epidemiologic and metabolomic approach improves CKD prediction. J. Am. Soc. Nephrol. **24**(8), 1330–1338 (2013)
73. Sharma, K., Karl, B., Mathew, A.V., Gangoiti, J.A., Wassel, C.L., Saito, R., et al.: Metabolomics reveals signature of mitochondrial dysfunction in diabetic kidney disease. J. Am. Soc. Nephrol. **24**(11), 1901–1912 (2013)
74. Vizcaíno, J.A., Csordas, A., Del-Toro, N., Dianes, J.A., Griss, J., Lavidas, I., et al.: 2016 update of the PRIDE database and its related tools. Nucleic Acids Res. **44**(22), 11033 (2016)
75. Siwy, J., Zürbig, P., Argiles, A., Beige, J., Haubitz, M., Jankowski, J., et al.: Noninvasive diagnosis of chronic kidney diseases using urinary proteome analysis. Nephrol. Dial. Transplant. **32**(12), 2079–2089 (2017)
76. Consortium, I.H.G.S.: Finishing the euchromatic sequence of the human genome. Nature. **431**(7011), 931–945 (2004)
77. Harrow, J., Frankish, A., Gonzalez, J.M., Tapanari, E., Diekhans, M., Kokocinski, F., et al.: GENCODE: the reference human genome annotation for The ENCODE Project. Genome Res. **22**(9), 1760–1774 (2012)
78. Buss, P.M.: Genomics and public health. Cad. Saude Publica. **18**(3), 554–555 (2002)
79. Groopman, E.E., Rasouly, H.M., Gharavi, A.G.: Genomic medicine for kidney disease. Nat. Rev. Nephrol. **14**(2), 83–104 (2018)
80. Lecamwasam, A., Sexton-Oates, A., Carmody, J., Ekinci, E.I., Dwyer, K.M., Saffery, R.: DNA methylation profiling of genomic DNA isolated from urine in diabetic chronic kidney disease: a pilot study. PLoS One. **13**(2), e0190280 (2018)
81. Sadar, S., Kaspate, D., Vyawahare, N.: Protective effect of L-glutamine against diabetes-induced nephropathy in experimental animal: role of KIM-1, NGAL, TGF-β1, and collagen-1. Ren Fail. **38**(9), 1483–1495 (2016)
82. Kaucsár, T., Godó, M., Révész, C., Kovács, M., Mócsai, A., Kiss, N., et al.: Urine/plasma neutrophil gelatinase associated lipocalin ratio is a sensitive and specific marker of subclinical acute kidney injury in mice. PLoS One. **11**(1), e0148043 (2016)
83. Luk, C.C., Chow, K.M., Kwok, J.S., Kwan, B.C., Chan, M.H., Lai, K.B., et al.: Urinary biomarkers for the prediction of reversibility in acute-on-chronic renal failure. Dis. Markers. **34**(3), 179–185 (2013)
84. Hemmingsen, C., Staun, M., Lewin, E., Egfjord, M., Olgaard, K.: Calcium metabolic changes and calbindin-D in experimental hypertension. J. Hypertens. **12**(8), 901–907 (1994)
85. Kumar, R., Schaefer, J., Grande, J.P., Roche, P.C.: Immunolocalization of calcitriol receptor, 24-hydroxylase cytochrome P-450, and calbindin D28k in human kidney. Am. J. Phys. **266**(3 Pt 2), F477–F485 (1994)
86. Guha, M., Heier, A., Price, S., Bielenstein, M., Caccese, R.G., Heathcote, D.I., et al.: Assessment of biomarkers of drug-induced kidney injury in cynomolgus monkeys treated with a triple reuptake inhibitor. Toxicol. Sci. **120**(2), 269–283 (2011)
87. Calero, M., Tokuda, T., Rostagno, A., Kumar, A., Zlokovic, B., Frangione, B., et al.: Functional and structural properties of lipid-associated apolipoprotein J (clusterin). Biochem. J. **344**(Pt 2), 375–383 (1999)
88. Wilson, M.R., Roeth, P.J., Easterbrook-Smith, S.B.: Clusterin enhances the formation of insoluble immune complexes. Biochem. Biophys. Res. Commun. **177**(3), 985–990 (1991)

89. Viard, I., Wehrli, P., Jornot, L., Bullani, R., Vechietti, J.L., Schifferli, J.A., et al.: Clusterin gene expression mediates resistance to apoptotic cell death induced by heat shock and oxidative stress. J. Invest. Dermatol. **112**(3), 290–296 (1999)
90. Leskov, K.S., Klokov, D.Y., Li, J., Kinsella, T.J., Boothman, D.A.: Synthesis and functional analyses of nuclear clusterin, a cell death protein. J. Biol. Chem. **278**(13), 11590–11600 (2003)
91. Xu, X., Wu, Z., Zhou, Q., Zhang, Y., Wu, D.: The role of determining the levels of serum collagen type IV in diagnosing early diabetic nephropathy. Ren. Fail. **24**(6), 747–753 (2002)
92. Perbal, B.: CCN proteins: multifunctional signalling regulators. Lancet. **363**(9402), 62–64 (2004)
93. Mosa, O., Skitek, M., Jerin, A.: Validity of Klotho, CYR61 and YKL-40 as ideal predictive biomarkers for acute kidney injury: review study. Sao Paulo Med. J. **135**(1), 57–65 (2017)
94. Sawai, K., Mukoyama, M., Mori, K., Kasahara, M., Koshikawa, M., Yokoi, H., et al.: Expression of CCN1 (CYR61) in developing, normal, and diseased human kidney. Am. J. Physiol. Renal Physiol. **293**(4), F1363–F1372 (2007)
95. Lai, C.F., Lin, S.L., Chiang, W.C., Chen, Y.M., Wu, V.C., Young, G.H., et al.: Blockade of cysteine-rich protein 61 attenuates renal inflammation and fibrosis after ischemic kidney injury. Am. J. Physiol. Renal Physiol. **307**(5), F581–F592 (2014)
96. Ulutas, O., Taskapan, M.C., Dogan, A., Baysal, T., Taskapan, H.: Vascular calcification is not related to serum fetuin-A and osteopontin levels in hemodialysis patients. Int. Urol. Nephrol. **50**(1), 137–142 (2018)
97. Matsui, I., Hamano, T., Mikami, S., Inoue, K., Shimomura, A., Nagasawa, Y., et al.: Retention of fetuin-A in renal tubular lumen protects the kidney from nephrocalcinosis in rats. Am. J. Physiol. Renal Physiol. **304**(6), F751–F760 (2013)
98. Westenfeld, R., Schäfer, C., Krüger, T., Haarmann, C., Schurgers, L.J., Reutelingsperger, C., et al.: Fetuin-A protects against atherosclerotic calcification in CKD. J. Am. Soc. Nephrol. **20**(6), 1264–1274 (2009)
99. Mehrsai, A., Guitynavard, F., Nikoobakht, M.R., Gooran, S., Ahmadi, A.: The relationship between serum and urinary Fetuin-A levels and kidney stone formation among kidney stone patients. Cent European J Urol. **70**(4), 394–399 (2017)
100. Stejskal, D., Karpisek, M., Vrtal, R., Student, V., Solichova, P., Fiala, R., et al.: Urine fetuin-A values in relation to the presence of urolithiasis. BJU Int. **101**(9), 1151–1154 (2008)
101. Andreucci, M., Faga, T., Pisani, A., Perticone, M., Michael, A.: The ischemic/nephrotoxic acute kidney injury and the use of renal biomarkers in clinical practice. Eur. J. Intern. Med. **39**, 1–8 (2017)
102. Shirakabe, A., Kobayashi, N., Hata, N., Shinada, T., Tomita, K., Tsurumi, M., et al.: The serum heart-type fatty acid-binding protein (HFABP) levels can be used to detect the presence of acute kidney injury on admission in patients admitted to the non-surgical intensive care unit. BMC Cardiovasc. Disord. **16**(1), 174 (2016)
103. Kokot, M., Biolik, G., Ziaja, D., Fojt, T., Kędzierski, L., Antoniak, K., et al.: Assessment of subclinical acute kidney injury after abdominal aortic aneurysm surgery using novel markers: L-FABP and H-FABP. Nefrologia. **34**(5), 628–636 (2014)
104. Ozturk, N., Aksoy, H., Aksoy, Y., Yildirim, A., Akcay, F., Yanmaz, V.: The low levels of circulating hepatocyte growth factor in nephrolithiasis cases: independent from gene polymorphism. Urolithiasis. **43**(5), 427–432 (2015)
105. Malatino, L.S., Cataliotti, A., Benedetto, F.A., Stancanelli, B., Bellanuova, I., Belluardo, P., et al.: Hepatocyte growth factor and left ventricular geometry in end-stage renal disease. Hypertension. **41**(1), 88–92 (2003)
106. Cho, W.Y., Lim, S.Y., Yang, J.H., Oh, S.W., Kim, M.G., Jo, S.K.: Urinary tissue inhibitor of metalloproteinase-2 and insulin-like growth factor-binding protein 7 as biomarkers of patients with established acute kidney injury. Korean J. Intern. Med. (2019)
107. Parikh, C.R., Jani, A., Melnikov, V.Y., Faubel, S., Edelstein, C.L.: Urinary interleukin-18 is a marker of human acute tubular necrosis. Am. J. Kidney Dis. **43**(3), 405–414 (2004)
108. Milas, O., Gadalean, F., Vlad, A., Dumitrascu, V., Velciov, S., Gluhovschi, C., et al.: Pro-inflammatory cytokines are associated with podocyte damage and proximal tubular dysfunction in the early stage of diabetic kidney disease in type 2 diabetes mellitus patients. J. Diabetes Complicat., 107479 (2019)
109. Ichimura, T., Asseldonk, E.J., Humphreys, B.D., Gunaratnam, L., Duffield, J.S., Bonventre, J.V.: Kidney injury molecule-1 is a phosphatidylserine receptor that confers a phagocytic phenotype on epithelial cells. J. Clin. Invest. **118**(5), 1657–1668 (2008)
110. Han, W.K., Bailly, V., Abichandani, R., Thadhani, R., Bonventre, J.V.: Kidney Injury Molecule-1 (KIM-1): a novel biomarker for human renal proximal tubule injury. Kidney Int. **62**(1), 237–244 (2002)
111. fibrosis, K.R.R.: KIM-1 expression links kidney injury with CKD in mice. Nat. Rev. Nephrol. **9**(11), 627 (2013)
112. Sabbisetti, V.S., Waikar, S.S., Antoine, D.J., Smiles, A., Wang, C., Ravisankar, A., et al.: Blood kidney injury molecule-1 is a biomarker of acute and chronic kidney injury and predicts progression to ESRD in type I diabetes. J. Am. Soc. Nephrol. **25**(10), 2177–2186 (2014)
113. Kamijo, A., Sugaya, T., Hikawa, A., Okada, M., Okumura, F., Yamanouchi, M., et al.: Urinary excretion of fatty acid-binding protein reflects stress overload on the proximal tubules. Am. J. Pathol. **165**(4), 1243–1255 (2004)
114. Yamamoto, T., Noiri, E., Ono, Y., Doi, K., Negishi, K., Kamijo, A., et al.: Renal L-type fatty acid--binding protein in acute ischemic injury. J. Am. Soc. Nephrol. **18**(11), 2894–2902 (2007)

115. Xu, Y., Xie, Y., Shao, X., Ni, Z., Mou, S.: L-FABP: a novel biomarker of kidney disease. Clin. Chim. Acta. **445**, 85–90 (2015)
116. Sheira, G., Noreldin, N., Tamer, A., Saad, M.: Urinary biomarker N-acetyl-β-D-glucosaminidase can predict severity of renal damage in diabetic nephropathy. J. Diabetes Metab. Disord. **14**, 4 (2015)
117. Bazzi, C., Petrini, C., Rizza, V., Arrigo, G., Napodano, P., Paparella, M., et al.: Urinary N-acetyl-beta-glucosaminidase excretion is a marker of tubular cell dysfunction and a predictor of outcome in primary glomerulonephritis. Nephrol. Dial. Transplant. **17**(11), 1890–1896 (2002)
118. Liangos, O., Perianayagam, M.C., Vaidya, V.S., Han, W.K., Wald, R., Tighiouart, H., et al.: Urinary N-acetyl-beta-(D)-glucosaminidase activity and kidney injury molecule-1 level are associated with adverse outcomes in acute renal failure. J. Am. Soc. Nephrol. **18**(3), 904–912 (2007)
119. Cirulli, V., Yebra, M.: Netrins: beyond the brain. Nat. Rev. Mol. Cell Biol. **8**(4), 296–306 (2007)
120. Ramesh, G., Krawczeski, C.D., Woo, J.G., Wang, Y., Devarajan, P.: Urinary netrin-1 is an early predictive biomarker of acute kidney injury after cardiac surgery. Clin. J. Am. Soc. Nephrol. **5**(3), 395–401 (2010)
121. Jayakumar, C., Nauta, F.L., Bakker, S.J., Bilo, H., Gansevoort, R.T., Johnson, M.H., et al.: Netrin-1, a urinary proximal tubular injury marker, is elevated early in the time course of human diabetes. J. Nephrol. **27**(2), 151–157 (2014)
122. Coles, M., Diercks, T., Muehlenweg, B., Bartsch, S., Zölzer, V., Tschesche, H., et al.: The solution structure and dynamics of human neutrophil gelatinase-associated lipocalin. J. Mol. Biol. **289**(1), 139–157 (1999)
123. Holmes, M.A., Paulsene, W., Jide, X., Ratledge, C., Strong, R.K.: Siderocalin (Lcn 2) also binds carboxymycobactins, potentially defending against mycobacterial infections through iron sequestration. Structure. **13**(1), 29–41 (2005)
124. Viau, A., El Karoui, K., Laouari, D., Burtin, M., Nguyen, C., Mori, K., et al.: Lipocalin 2 is essential for chronic kidney disease progression in mice and humans. J. Clin. Invest. **120**(11), 4065–4076 (2010)
125. Koyner, J.L., Vaidya, V.S., Bennett, M.R., Ma, Q., Worcester, E., Akhter, S.A., et al.: Urinary biomarkers in the clinical prognosis and early detection of acute kidney injury. Clin. J. Am. Soc. Nephrol. **5**(12), 2154–2165 (2010)
126. Lau, W.K., Blute, M.L., Weaver, A.L., Torres, V.E., Zincke, H.: Matched comparison of radical nephrectomy vs nephron-sparing surgery in patients with unilateral renal cell carcinoma and a normal contralateral kidney. Mayo Clin. Proc. **75**(12), 1236–1242 (2000)
127. Fassett, R.G., Venuthurupalli, S.K., Gobe, G.C., Coombes, J.S., Cooper, M.A., Hoy, W.E.: Biomarkers in chronic kidney disease: a review. Kidney Int. **80**(8), 806–821 (2011)
128. du Cheyron, D., Daubin, C., Poggioli, J., Ramakers, M., Houillier, P., Charbonneau, P., et al.: Urinary measurement of Na+/H+ exchanger isoform 3 (NHE3) protein as new marker of tubule injury in critically ill patients with ARF. Am. J. Kidney Dis. **42**(3), 497–506 (2003)
129. Kaleta, B.: The role of osteopontin in kidney diseases. Inflamm. Res. **68**(2), 93–102 (2019)
130. Kaleta, B., Krata, N., Zagożdżon, R., Mucha, K.: Osteopontin gene polymorphism and urinary OPN excretion in patients with immunoglobulin a nephropathy. Cell. **8**(6) (2019)
131. Wasung, M.E., Chawla, L.S., Madero, M.: Biomarkers of renal function, which and when? Clin. Chim. Acta. **438**, 350–357 (2015)
132. Cho, J.H., Ryu, H.M., Oh, E.J., Yook, J.M., Ahn, J.S., Jung, H.Y., et al.: Alpha1-antitrypsin attenuates renal fibrosis by inhibiting TGF-β1-induced epithelial mesenchymal transition. PLoS One. **11**(9), e0162186 (2016)
133. Weber, M.H., Verwiebe, R.: Alpha 1-microglobulin (protein HC): features of a promising indicator of proximal tubular dysfunction. Eur. J. Clin. Chem. Clin. Biochem. **30**(10), 683–691 (1992)
134. Jotwani, V., Scherzer, R., Abraham, A., Estrella, M.M., Bennett, M., Cohen, M.H., et al.: Association of urine α1-microglobulin with kidney function decline and mortality in HIV-infected women. Clin. J. Am. Soc. Nephrol. **10**(1), 63–73 (2015)

Urinary Extracellular Vesicles Magic Particles for Biomarker Discovery

3

Karina Barreiro, Tobias B. Huber, and Harry Holthofer

Abstract

Extracellular vesicles (EV) are small membrane-coated structures secreted by all cells of the body and can be detected in all bodily fluids, including urine. EV contents (e.g. proteins and distinct RNA classes) reflect the physiological state of their cells of origin, offering a new source of biomarkers. Accordingly, urinary Extracellular Vesicles (uEVs) are emerging as a source for early biomarkers of kidney damage and beyond, holding the potential to replace the conventional invasive techniques including kidney biopsy. However, the lack of standardization and sample collection and isolation methods, and the influence of factors such as inter- and intra-individual variability create difficulties in interpreting current results. Here we review recent results and reported uses of especially urinary EVs and also pinpoint approaches to be considered when designing experiments.

Keywords

Urinary extracellular vesicles · Exosomes · Biomarkers

3.1 Introduction

While the history of urinary Extracellular Vesicles (uEV) is relatively young, knowledge of their biologic roles appear to grossly follow those of vesicles found in other bodily fluids.

With the already impressive impact list of EVs, it is fair to state that these abundant vesicular structures once considered as handy waste packages to be excreted out of the body, now appear to open new understanding to many key biological phenomena. These include, but not limited to, general cell-to-cell communication and regulation of immune reaction and extends to roles in spread of cancer cells to precise modulation of tissue -and cell-type specific functions.

Detailed information of molecular mechanisms mediated by EVs and their "holistic" understanding in distinct tissue functions is rapidly emerging. Notably, many of these mecha-

K. Barreiro
Institute of Molecular Medicine Finland, University of Helsinki, Helsinki, Finland
e-mail: karina.barreiro@helsinki.fi

T. B. Huber
Department of Medicine, University Medical Center Hamburg-Eppendorf, Hamburg, Germany
e-mail: t.huber@uke.de

H. Holthofer (✉)
Institute of Molecular Medicine Finland, University of Helsinki, Helsinki, Finland

Department of Medicine, University Medical Center Hamburg-Eppendorf, Hamburg, Germany
e-mail: harry.holthofer@helsinki.fi; h.holthoefer@uke.de

H. M. Baptista Carreira dos Santos (ed.), *Translational Urinomics*, Advances in Experimental Medicine and Biology 1306, https://doi.org/10.1007/978-3-030-63908-2_3

nisms have now been established and appear to parallel generic EV functions in different biological fluids. Interestingly, many recent reports have highlighted the potential of customized EVs as targeted future "magic bullets" in precision medicine to enable advanced personalized and tissue targeted treatments. EVs are lucrative sources of personalized disease biomarkers, snapshots of cellular and tissue pathophysiology. These snapshots provide unforeseen accuracy to monitor distinct functions and, more importantly, to highlight new understanding of basic cellular biological events.

Reproducible use of extracellular vesicles in academic or applied research for clinically relevant problems still needs stringent standardization at many levels including sample collection, storage, EV analytics and downstream applications. In spite of valuable standardization guidelines already achieved, with particular and continued efforts by the International Society for Extracellular Vesicles [1], many important steps remain to be achieved and strict adherence to published guidelines enforced. During the rapid expansion of EV applications in a multitude of areas and from an ever increasing number of laboratories, the sheer credibility and reproducibility of results will need much stronger attention. Before EVs can be adopted in wide general use, especially in the biomarker applications, several fundamental limitations and controversies still exist.

Here we review recent results and reported uses of especially urinary EVs and also pinpoint approaches to be considered when designing experiments.

3.2 Urinary EV Biology

Extracellular vesicles are well characterized to their main categories of exosomes, microvesicles and apoptotic bodies (Fig. 3.1). This division is somewhat arbitrary and mostly considers the biogenesis and size of vesicles, without going into other physicochemical features nor into their precise biochemical characteristics. These basic EV properties have been well described several recent excellent reviews [2–6].

The milestone report of urinary EVs by Pisitkun et al. 2004 translated the physicochemical features of EVs also into urine, including their size, shape and contents. Thus, it appears that uEVs show the similar variety of protein, RNA and lipid content as EVs from other biological sources. This suggests but does not prove that EV functions in the distinct sources are identical.

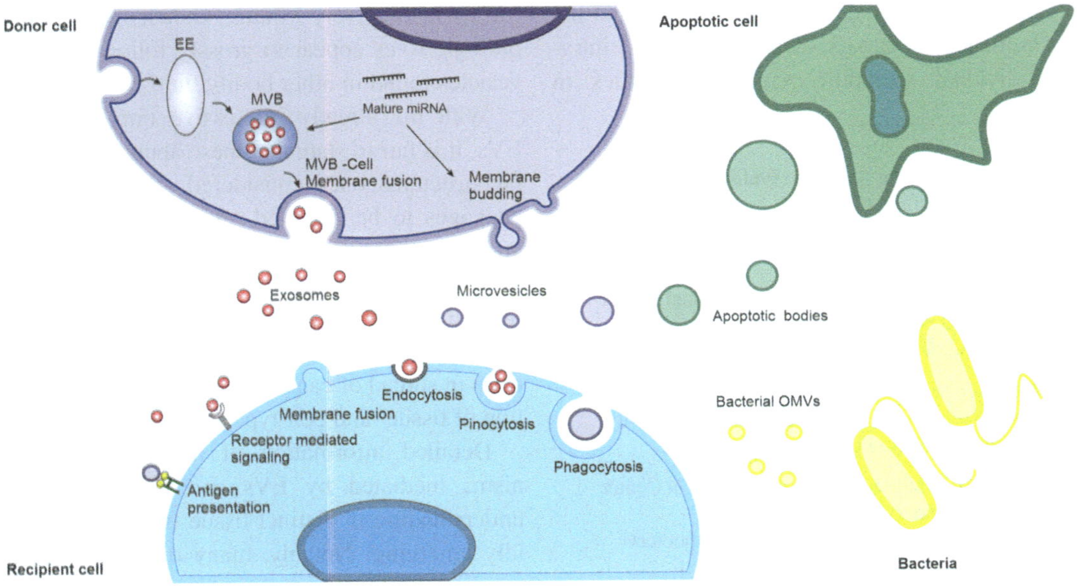

Fig. 3.1 Schematic representation of EV sources and biology. *EE* early endosome, *MVB* multivesicular bodies, *OMVs* outer membrane vesicles

Traditional view tells that urine is the vehicle to excrete metabolic waste products from not only the kidney tissue but, via glomerular filtration and tubular secretion, from the whole body. It is becoming increasingly evident, mostly by uEV studies, that urine is far beyond simple means to excrete waste from the body but indeed is a complex, poorly understood mixture of different biologically active molecular species. Notably, urinary contents dynamically reflect metabolic extremes within an individual, e.g. after strong physical exercise leading to dehydration [7]. These extremes result in well-established changes in urinary osmolality as well as in some key urinary "normal contents". While uEV changes could reflect individual physiological adaptation to stress situations at the organ level, it is also evident that such changes are similarly seen following shift in e.g. dietary patterns, disease states and upon introduction of new medication. This means that uEV content reflects more the total body reaction and implies that all uEV clinical studies should also contain comprehensive reporting of clinical chemistry values as well as activity, diet, and, particularly, use of medications.

With the uEV contents of specific proteins beyond the kidney, it is conceivable that uEVs indeed reflect systemic rather than merely local tissue level physiology. How EVs from circulation get access to urine remains an open question but could involve transglomerular passage through the glomerular filtration barrier (GFB) or active secretion by tubular epithelial cells.

Normal urine contains cell type-specific proteins from the glomerular filtration barrier, particularly the glomerular visceral epithelial cells (podocytes) [8], and electron microscopy of glomeruli also show abundance of vesicles in this site [9]. Final proof visualizing EVs passaging the GFB remain to be achieved.

EV contents in general include specific functional, structural and metabolic proteins, lipids and, particularly, a rich content of RNA species (mi/messenger/lncRNA) and DNA [4, 10]. uEVs show a closely similar repertoire of molecules [6]. While their respective physiologic significance remains still poorly studied, it is tempting to speculate that these active molecules continue to have their respective functions also in urine. In this respect, EV surface molecules could serve as address codes to target their active contents to specific downstream sites [8]. However, further consensus needs to be reached on the methods to isolate and characterize the EV surface contents as passive adsorption of a variety of molecules in this site is possible [11, 12]. After targeting and adhesion, EVs may be taken up by a variety of well described mechanisms [13–15] to subsequently induce functional changes [5].

It is interesting to note that diet [16], use of medication, exercise and other still poorly understood factors [17–19] may cause intra-individual changes in vesicle content as detected by proteomics or RNomics. Our own results indicate significant intra-individual variation in first morning urine uEV size, class and especially contents distribution during daily monitoring over 3 months of first morning urine voids (Holthofer et al., unpublished).

Our recent results have shown distinct gender and ethnic differences in uEV contents of healthy subjects, which may be only partly explained by dietary differences (Xu et al., unpublished). These results emphasize the crucial importance of recording comprehensive phenotypic data of subjects under study to understand the dynamic changes in uEV signatures and factors behind. With this complexity in mind, our approach to understand biological roles of uEVs has consisted of using three pronged approach (see Fig. 3.2) and, especially, established animal models of diseases under study. Notably, this approach as a first step eliminates differences based on e.g. gender, genetic heterogeneity, diet, medication, and exercise (Xu et al., unpublished). It is also interesting to note that our preliminary studies with human subjects show rapid changes in distinct uEV categories upon medication: when type II (adult onset) diabetics started with SGLT1 inhibitor or miconatsole, distinct intra-individual changes in miRNA were observed (Barreiro et al., unpublished). This shows the rapid, dynamic changes in cellular physiology that can be recapitulated as cellular content snapshots in the form of uEV changes.

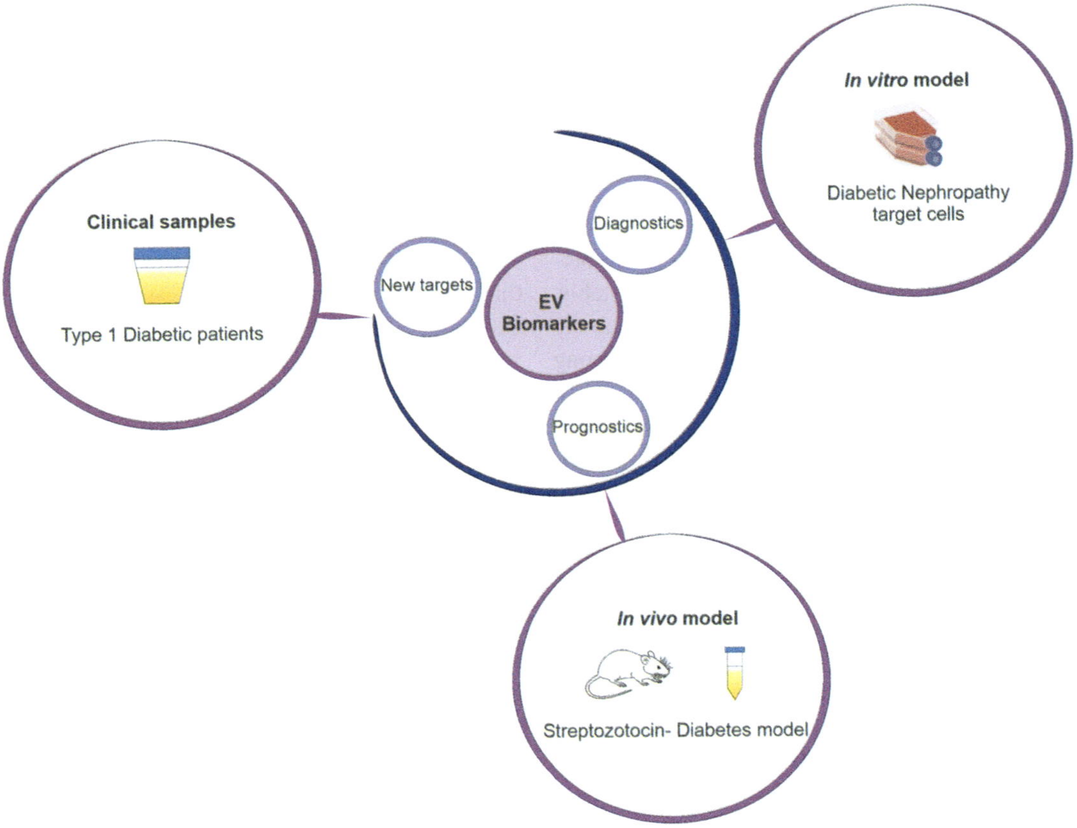

Fig. 3.2 Overarching experimental setups to verify uEV biomarkers of diabetic kidney disease targets. *EV* extracellular vesicles

Taken together, there are currently many unknown individual factors which may explain the observed complexity of uEV patterns (as well as EVs from other bodily sources). It is also evident that significant advances need to be achieved to understand the abundance of factors influencing uEV contents before they can be fully introduced as biomarkers or tools to precision medicine.

3.3 Urinary Extracellular Vesicle Isolation

Currently, numerous methods are used for extracellular vesicle isolation e.g. ultracentrifugation [6], filtration [20], precipitation [21] and hydrostatic filtration dialysis [22] based techniques, or combination of these. Isolation methods make use of several of the extracellular vesicle properties e.g. size, density and solubility and this is why different subpopulations of extracellular vesicles are enriched by different methods. The method or combination of methods to be applied usually depends on the starting material and the volume [23]. However, there is presently no solid standardization approaches of protocols and, in many cases, methods applied are reported loosely to confound result interpretation and repeatability. Details on the isolation methods, advantages and disadvantages such as processing times, costs, ease of implementation, and co-isolation of contaminants are well described in recent detailed reviews [8, 24–27]. For reporting a new method, a modification of an existing method or, when applying a method for the first time, we strongly recommend strict adherence to MISEV2018 guideline details [1].

3.4 Inter- and Intra- Individual Variability in EVs Samples

Few studies have assessed systematically the intra- and inter- individual variation of EV contents, and most of them focus on results based on proteomics findings.

Accordingly, in many proteomics based studies, it has been reported that inter- and intra- individual variation is lower in EVs compared to crude urine [28]. In addition, inter-individual variation was described to be higher than intra-individual variation in urinary EVs [29]. In contrast, both inter- and intra- variation for cell culture EVs proteins were reported to be low [30].

Interestingly, the opposite was described for miRNA urinary EV content (inter-individual variation was lower than intra- individual variation) [31] and less variability for miRNA isolated from plasma derived EVs [32] was observed.

The differences regarding inter and intra- individual variation in these reports may be due to the use of different biological fluids, differences in pre-analytical variables and to the diversity of EV isolation methods used. Thus, for biomarkers research it is of key importance to determine the variation for each sample type, downstream application and, vesicle isolation method in order to calculate a solid sample number [33].

Age, gender, and race differences have also been reported to affect the size of the vesicles and the protein content [34, 35]. Thus, these factors should be considered for a well-balanced experimental design.

3.5 Example of Experimental Setups Relevant in uEV Studies

To avoid oversimplification and misinterpretation of uEV results seen especially in urines from diabetic patients, our experimental setup has built on utilization of three levels of meticulously monitored EV parameters as follows:

1. EVs from *in vitro* cultured diabetes target cells
2. uEVs from established experimental model of diabetes in the rat
3. uEVs from human diabetic (type 1 diabetes) patient urines

This comprehensive approach to study uEVs in Diabetic Kidney Disease (DKD), starting from in vitro and in vivo models to human samples, aims to identify better tools and predictive marker candidates for early DKD diagnostics and disease management.

For the in vitro studies (Figs. 3.3 and 3.4) our hypothesis was that the use of known diabetes target cells and their EV secretion patterns upon diabetic conditions gives the most simple and targeted information of respective EV secretion responses. This approach allows to study effects of a variety of modifications, e.g. variations of insulin, glucose or established pharmacologic manipulation with standardized harvesting of EV response, respectively. Thus, we used conditioned culture media from all three resident glomerular cell types, the kidney targets of diabetes: podocytes, mesangial and endothelial cells with the adjacent proximal tubular cells. According to the study protocol, cells were grown with or without diabetes -mimicking conditions (collaboration with Prof Richard Coward's lab, Bristol University, UK). EVs from cell culture media were isolated by Hydrostatic Filtration dialysis [36]. Rigorous quality control of the isolated vesicles was done by Western blot, nanoparticle analysis and electron microscopy. Samples were processed for small and long RNA sequencing, and quantitative proteomics. With this data, we seek to define a panel of dysregulated miRNAs, mRNAs and proteins (by cell type) in these in vitro conditions. All data layers are being integrated with multiomics approach to reveal novel candidate pathways involved in insulin resistance and other parameters. Preliminary results of the study from podocytes are shown.

3.6 uEV Omics Studies for Biomarker Research

The application of omics techniques to EV samples has increased remarkably as a consequence of general interest into EV for biomarker research. This is clearly reflected in the amount

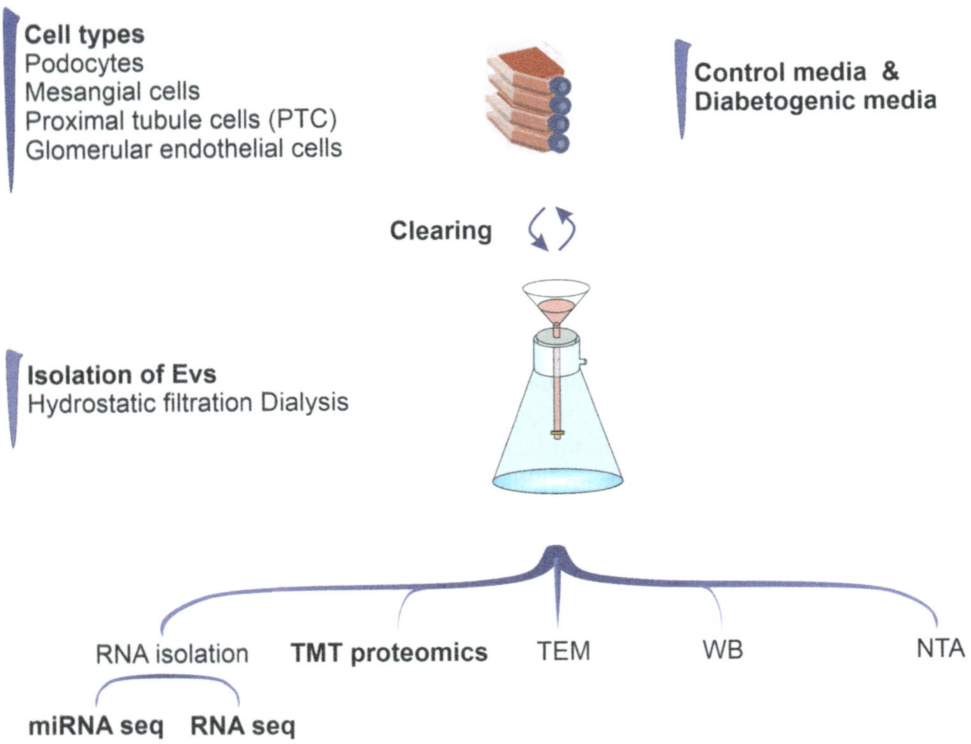

Fig. 3.3 Experimental workflow for EVs derived from cell culture lines to study diabetic kidney disease. *EVs* extracellular vesicles, *TMT proteomics* tandem mass tag proteomics, *TEM* transmission electron microscopy, *WB* Western blot, *NTA* nanoparticle tracking analysis

of publications involving EV proteomics, transcriptomics, lipidomics, or metabolomics, which are thoroughly documented in recent reviews [37–41]. Even if these techniques offer exciting possibilities to profile EV contents precisely, it is important to consider the technical challenges that accompany them in order to design experiments accordingly [42]. Table 3.1 shows examples of recent studies involving specifically urinary extracellular vesicles. As evident from Table 3.1, the use of different approaches to isolate EV, workflows for isolation of features under study (e.g. RNA, proteins), and normalization protocols vary from study to study. The lack of standardization in general and only a few pilot studies available to compare particular omics approaches using same sample sets affects interpretation on what is the preferential approach to study uEV.

All omics approaches can be considered ultrasensitive. This fact, especially in transcriptomics approaches emphasizes the crucial importance of selecting optimized uEV isolation methods and their critical application at all steps. Poor early quality measures will automatically result in poor data quality. Reports have shown that different uEV isolation methods result in dissimilar miRNA containing EV populations and/or even co-isolate non-EV miRNA [43–45]. However, most of the comparisons have thus far focused in uEV miRNA analysis. Thus, more studies are needed to reach a clear picture on the effect of early EV isolation steps on RNA type outcomes.

Furthermore, the library preparation protocol applied has major effects on sequencing results as e.g. (i) small vs total RNA library approaches affect RNA biotype distribution [40], (ii) poly(A) library preparation approach does not perform well with partially degraded RNAs. Thus, integ-

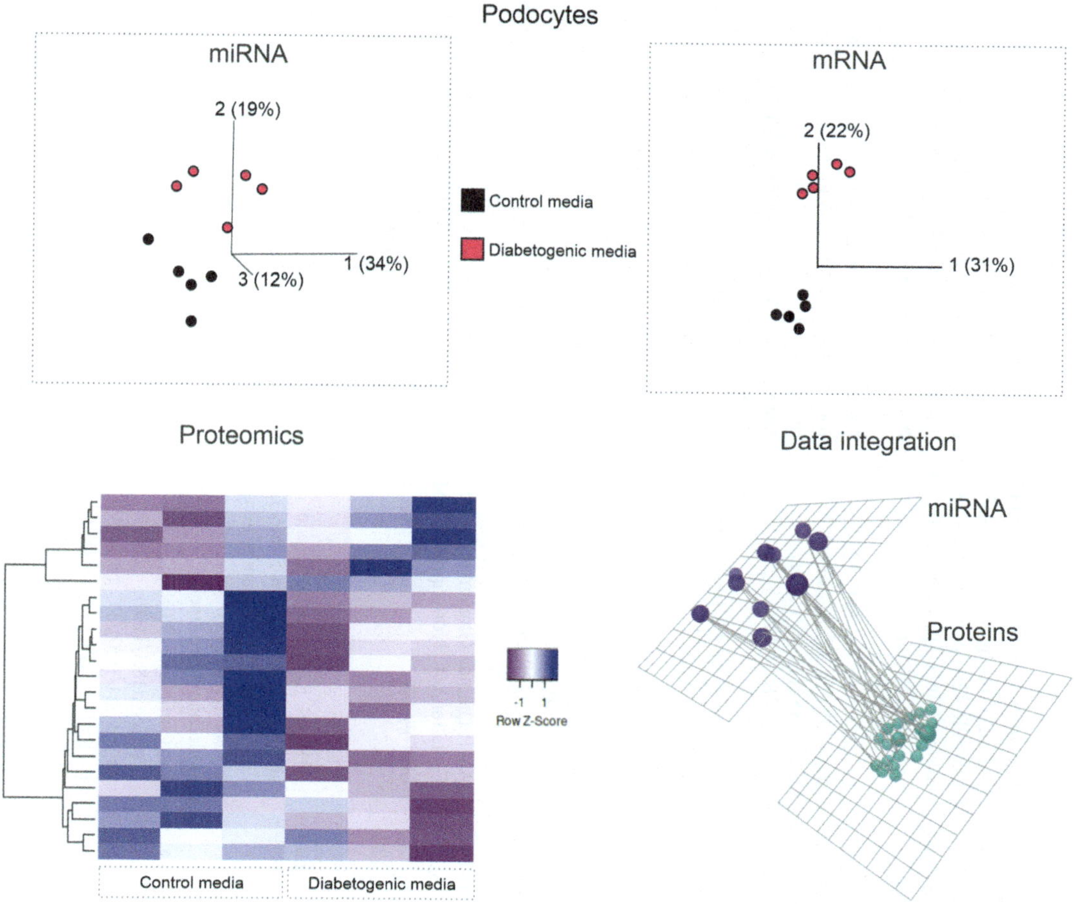

Fig. 3.4 Representative results obtained for the in vitro model of diabetic nephropathy. PCA of small and long RNA sequencing (**a**, **b**), expression heatmap of proteomics data [58] (**c**) and multiomics integration of miRNA and Proteomics data [59] (**d**)

rity of EV samples RNA has to be evaluated before choosing this kind of library, (iii) Library Kits that employ universal adapters with fixed sequences generate biases in small RNA sequencing. As reported by Srinivasan et al. [46], the kits that use adapters with degenerate bases reduce biases for small RNA sequencing.

Normalization of quantitative data is also a possible confounding factor. Whether it is better to normalize per starting volume of urine, creatinine or EV related values (e.g. particle number, presence of CD9) is not clear and remains an open debate [47]. MISEV 2018 guidelines acknowledge the lack of agreement on normalization strategy and calls for more studies on the topic.

Validation of sequencing results is usually done through qPCR. However, for example the lack of a reference miRNA and biases introduced by the library preparation itself could difficult validation of the respective miRNA targets [48]. As housekeeping genes used normally may not work properly for EV sequencing data, ISEV recommends to use multiple reference genes to normalize qPCR data [49]. For more information on EV RNA analysis, methodologies and biases, we recommend an ISEV position review by Mateescu et al. [50].

Table 3.1 Selected recent urinary extracellular vesicles studies involving omics approaches to identify biomarkers

Omics analytics	Disease	uEV Isolation Method	Workflow	Examples of features found in uEVs	Comments	Ref
Transcriptomics	Type 1 diabetes, Diabetic nephropathy	Size exclusion chromatography	RNA isolation using miRNeasy Micro kit, in-house small-RNAseq library preparation and sequencing using NextSeq sequencer.	hsa-miR-30a-3p, hsa-miR-204-5p, hsa-miR-429-3p, hsa-miR-141-3p, hsa-miR-365a-3p, hsa-miR-30a-5p, hsa-miR-200c-3p[a]	EV isolated from 250 μl of urine was used for RNA isolation. Validation by TaqMan Advanced miRNA assays using hsa-miR-16-5p for normalization.	Ghai et al. [60]
Transcriptomics	Chronic kidney disease	Ultacentrifugation + DTT	RNA isolated using Tri-reagent. Ion Total RNA-Seq Kit v2 used for library preparation and sequenced using Ion Proton sequencer.	hsa-let-7c-5p, hsa-miR-222-3p, hsa-miR-27a-3p, hsa-miR-27b-3p, EAF1-AS1, PCBP1-AS1, RP11-138E2.1_linc[b]	EV isolated from 50 ml of urine was used for RNA isolation. Authors developed an in silico pipeline to analyze the small ncRNA transcriptome "ncRNASeqScan".	Khurana et al. [61]
Proteomics	Parkinson's Disease	Ultracentrifugation	In gel digestion, HPLC-MS.	SNAP23, ES8L2, CPM, FRK, STK24, CALB1[c]	uEV pellets from ~40 mL of urine were used for protein extraction. Then, equal amount of proteins (10 ug) were further processed. Validation by immunoblot using heat-shock protein 70 for normalization.	Wang et al. [29]
Proteomics	Polycystic Kidney Disease	Ultracentrifugation + DTT	uEV identification cohort: quantitative dimethyl labeling LC-MS/MS.	DSP, EVPL, CFB, TTYH3, VTCN1, GPRC5A[d]	Validation by immunoblot using CD9 for normalization.	Salih et al. [62]
Metabolomics	Prostate cancer	Ultracentrifugation	Metabolites extracted from EV employing Acidic acetonitrile solvent. UHPLC-MS-MS	AMP, Gamma-Glutamylcysteine, Glutatione, 4-Hydroxiproline, IMP, Spermidine[e]	Approximately > = 1 x 10^10 EV was required for an EV metabolome profile. For data normalization, various approaches were explored. Best results obtained with EV-derived factors and analysis of metabolites ratios.	Puhka et al. [63]
Metabolomics	Prostate cancer	Ultracentrifugation	Methanol–chloroform solvent for metabolite extraction from uEV. UHPLC-MS.	DHEAS, isomer androsterone sulfate 1, isomer pregn-5-ene-3,20-diol sulphate, isomer androsterone sulfate 2, AC(14:1n-x), Dodecanoylcarnitine[f]	EV isolated from 50 ml of urine were used to isolate metabolites.	Clos-Garcia et al. [64]
Lipidomics	Prostate cancer	Ultracentrifugation	Lipids extracted from urinary exosomes using a modified Folch lipid extraction procedure. LC-MS/MS.	HexCer(d18:1/12:0), LAcCer(d18:1/16:0), Gb3(d18:1/16:0), PC 16:0-18:2, PE P-18:0/18:2 (PE O-18:1/18:2), PS 18:0-18:2[g]	Equal amount of uEV proteins (20 or 4 ug) were used as a reference for lipid extraction.	Skotland et al. [65]

[a]miRNAs enriched in uEV versus urine from Type 1 diabetic patients. [b]Up-regulated ncRNAs in uEV from chronic kidney disease (stages I, II, III and IV) versus healthy controls. [c]Proteins enriched in uEV from Parkinson disease patients vs control in the discovery cohort. [d]Proteins abundant in uEV of patients with autosomal dominant polycystic kidney disease. [e]Metabolites enriched in uEV vs urine. [f]Metabolites enriched in uEV from prostate cancer patients. [g]Lipids enriched in uEV from prostate cancer patients [29, 60–65]

3.7 Future of uEV Studies

Urine remains an underutilized diagnostic matrix, which can be easily and painlessly collected in large quantities, repeatedly and processed easily for further analytics. Factors limiting its diagnostic consist of dynamic contents together with constituents causing serious artifacts in downstream analytics. For these reasons, rigorous standardization of urine is needed and methods are now available for successful use of urine. Notably, to solve these challenges we recently developed a method combining equalization of urinary electrolytes while neutralizing effects of pigments and manage Tamm-Horsfall [51]. Interestingly, the method developed simultaneously suits perfectly for unbiased collection of uEVs to release the full analytical power of the uEVs. It is notable, that uEVs contain 5–20 times the contents of proteins, RNA and lipids as compared to crude urine and, furthermore, without many known artifacts associated with the use of crude urine.

With the wealth of data being increasingly published on the biology, derivation, association to disease and future therapeutic aspects of EVs, it is now reasonable to expect that the uEVs will show their power as future analytic and source for biomarkers valuable for precision medicine.

To process the large number of samples needed to define reliable biomarkers, automated methods to isolate efficiently extracellular vesicles are needed. This is especially important to reduce the processing time; to reduce the number of steps needed to process samples (to reduce batch effects). Reducing these factors to essential ones, addition of technical replicates and follow up samples in large studies would be achievable in reasonable times. However, performance of commercially available kits (better for high-throughput) is variable when compared to more traditional and well characterized isolation methods [52, 53]. More studies are needed to compare isolation methods including control and disease samples and their use in downstream applications. In addition to the commercially available kits new approaches are emerging to automate and standardize the isolation process [54].

With the increasing interest in EVs as source of novel biomarkers, the need to develop devices that could both isolate and detect markers (Lab-on-a-chip) has grown substantially. Promising devices have already been designed, based e.g. in diverse microfluidic strategies to isolate successfully extracellular vesicles as well as detection of their RNA and protein contents and the combination of both on single device [55, 56]. These approaches provide an interesting future solution for quick EV biomarker detection in clinical settings.

Together with the innovation in isolation methods, exiting advances were made in single EV detection in e.g. flow cytometry, which could be applied for accurate biomarker detection [57].

Acknowledgements Research lines reported here are based on fruitful collaborative efforts and are gratefully acknowledged as follows:

Prof Richard Coward's team (Bristol University, UK, in-vitro models);

Prof Leif Groop's team (University of Helsinki, Finland, Clinical samples);

Dr. Carol Forsblom (The Finnish Diabetic Nephropathy Study, Finland, Clinical Samples;);

Prof Riitta Lassila (University of Helsinki, Finland, in vivo experimental model);

And Dr. Denis Delic, Dr. German Leparc, Marcel Rosler (Boehringer Ingelheim Pharma GmbH & Co. KG, Germany, short and long RNA sequencing)This study was supported by grants from the Paulo Foundation of Finland and NovoNordisk Foundation.

TBH was supported by the DFG (CRC1192, HU 1016/8-2, HU 1016/11-1, HU 1016/12-1), by the BMBF (STOP-FSGS-01GM1518C), and by the European Research Council-ERC (grant 616891).

This project has received funding from the Innovative Medicines Initiative 2 Joint Undertaking under grant agreement no. 115974 (BEAt-DKD; T.B.H., H.H.). This Joint Undertaking receives support from the European Union's Horizon 2020 research and innovation programme and EFPIA and by the H2020-IMI2 consortium BEAt-DKD.

References

1. Thery, C., Witwer, K.W., Aikawa, E., Alcaraz, M.J., Anderson, J.D., Andriantsitohaina, R., et al.: Minimal information for studies of extracellular vesicles 2018 (MISEV2018): a position statement of the International Society for Extracellular Vesicles and update of the MISEV2014 guidelines. J. Extracell. Vesicles. **7**(1), 1535750 (2018)
2. Kalluri, R., LeBleu, V.S.: The biology, function, and biomedical applications of exosomes. Science (New York, NY). **367**(6478) (2020)
3. van Niel, G., D'Angelo, G., Raposo, G.: Shedding light on the cell biology of extracellular vesicles. Nat. Rev. Mol. Cell Biol. **19**(4), 213–228 (2018)
4. Yanez-Mo, M., Siljander, P.R., Andreu, Z., Zavec, A.B., Borras, F.E., Buzas, E.I., et al.: Biological properties of extracellular vesicles and their physiological functions. J. Extracell. Vesicles. **4**, 27066 (2015)
5. Maas, S.L.N., Breakefield, X.O., Weaver, A.M.: Extracellular vesicles: unique intercellular delivery vehicles. Trends Cell. Biol. **27**(3), 172–188 (2017)
6. Pisitkun, T., Shen, R.F., Knepper, M.A.: Identification and proteomic profiling of exosomes in human urine. Proc. Natl. Acad. Sci. U. S. A. **101**(36), 13368–13373 (2004)
7. Maughan, R.J., Shirreffs, S.M.: Dehydration and rehydration in competative sport. Scand. J. Med. Sci. Sports. **20**(Suppl 3), 40–47 (2010)
8. Musante, L., Tataruch, D.E., Holthofer, H.: Use and isolation of urinary exosomes as biomarkers for diabetic nephropathy. Front. Endocrinol. **5**, 149 (2014)
9. Rastaldi, M.P., Armelloni, S., Berra, S., Li, M., Pesaresi, M., Poczewski, H., et al.: Glomerular podocytes possess the synaptic vesicle molecule Rab3A and its specific effector rabphilin-3a. Am. J. Pathol. **163**(3), 889–899 (2003)
10. Huttenhofer, A., Mayer, G.: Circulating miRNAs as biomarkers of kidney disease. Clin. Kidney J. **10**(1), 27–29 (2017)
11. Miranda, K.C., Bond, D.T., McKee, M., Skog, J., Paunescu, T.G., Da Silva, N., et al.: Nucleic acids within urinary exosomes/microvesicles are potential biomarkers for renal disease. Kidney Int. **78**(2), 191–199 (2010)
12. Bryzgunova, O.E., Zaripov, M.M., Skvortsova, T.E., Lekchnov, E.A., Grigor'eva, A.E., Zaporozhchenko, I.A., et al.: Comparative study of extracellular vesicles from the urine of healthy individuals and prostate cancer patients. PLoS One. **11**(6), e0157566 (2016)
13. Horibe, S., Tanahashi, T., Kawauchi, S., Murakami, Y., Rikitake, Y.: Mechanism of recipient cell-dependent differences in exosome uptake. BMC Cancer. **18**(1), 47 (2018)
14. Mathieu, M., Martin-Jaular, L., Lavieu, G., Thery, C.: Specificities of secretion and uptake of exosomes and other extracellular vesicles for cell-to-cell communication. Nat. Cell Biol. **21**(1), 9–17 (2019)
15. Mulcahy, L.A., Pink, R.C., Carter, D.R.: Routes and mechanisms of extracellular vesicle uptake. J. Extracell. Vesicles. **3** (2014)
16. Eitan, E., Tosti, V., Suire, C.N., Cava, E., Berkowitz, S., Bertozzi, B., et al.: In a randomized trial in prostate cancer patients, dietary protein restriction modifies markers of leptin and insulin signaling in plasma extracellular vesicles. Aging Cell. **16**(6), 1430–1433 (2017)
17. Rigamonti, A.E., Bollati, V., Pergoli, L., Iodice, S., De Col, A., Tamini, S., et al.: Effects of an acute bout of exercise on circulating extracellular vesicles: tissue-, sex-, and BMI-related differences. Int. J. Obes. **2019** (2005)
18. Fruhbeis, C., Helmig, S., Tug, S., Simon, P., Kramer-Albers, E.M.: Physical exercise induces rapid release of small extracellular vesicles into the circulation. J. Extracell. Vesicles. **4**, 28239 (2015)
19. Zachar, R., Jensen, B.L., Svenningsen, P.: Dietary Na(+) intake in healthy humans changes the urine extracellular vesicle prostasin abundance while the vesicle excretion rate, NCC, and ENaC are not altered. Am. J. Physiol. Renal Physiol. **317**(6), F1612–F1f22 (2019)
20. Cheruvanky, A., Zhou, H., Pisitkun, T., Kopp, J.B., Knepper, M.A., Yuen, P.S., et al.: Rapid isolation of urinary exosomal biomarkers using a nanomembrane ultrafiltration concentrator. Am. J. Physiol. Renal Physiol. **292**(5), F1657–F1661 (2007)
21. Rider, M.A., Hurwitz, S.N., Meckes Jr., D.G.: ExtraPEG: a polyethylene glycol-based method for enrichment of extracellular vesicles. Sci. Rep. **6**, 23978 (2016)
22. Musante, L., Tataruch, D., Gu, D., Benito-Martin, A., Calzaferri, G., Aherne, S., et al.: A simplified method to recover urinary vesicles for clinical applications, and sample banking. Sci. Rep. **4**, 7532 (2014)
23. Gardiner, C., Di Vizio, D., Sahoo, S., Théry, C., Witwer, K.W., Wauben, M., et al.: Techniques used for the isolation and characterization of extracellular vesicles: results of a worldwide survey. J. Extracell. Vesicles. **5**, 32945 (2016)
24. Konoshenko, M.Y., Lekchnov, E.A., Vlassov, A.V., Laktionov, P.P.: Isolation of extracellular vesicles: general methodologies and latest trends. Biomed. Res. Int. **2018**, 8545347 (2018)
25. Barreiro, K., Holthofer, H.: Urinary extracellular vesicles. A promising shortcut to novel biomarker discoveries. Cell Tissue Res. **369**(1), 217–227 (2017)
26. Merchant, M.L., Rood, I.M., Deegens, J.K.J., Klein, J.B.: Isolation and characterization of urinary extracellular vesicles: implications for biomarker discovery. Nat. Rev. Nephrol. **13**(12), 731–749 (2017)
27. van der Pol, E., Boing, A.N., Gool, E.L., Nieuwland, R.: Recent developments in the nomenclature, presence, isolation, detection and clinical impact of extracellular vesicles. J. Throm. Haem. **14**(1), 48–56 (2016)

28. Lee, J., McKinney, K.Q., Pavlopoulos, A.J., Niu, M., Kang, J.W., Oh, J.W., et al.: Altered proteome of extracellular vesicles derived from bladder cancer patients urine. Mol. Cells. **41**(3), 179–187 (2018)
29. Wang, S., Kojima, K., Mobley, J.A., West, A.B.: Proteomic analysis of urinary extracellular vesicles reveal biomarkers for neurologic disease. EBioMedicine. **45**, 351–361 (2019)
30. Tiruvayipati, S., Wolfgeher, D., Yue, M., Duan, F., Andrade, J., Jiang, H., et al.: Variability in protein cargo detection in technical and biological replicates of exosome-enriched extracellular vesicles. PLoS One. **15**(3), e0228871 (2020)
31. Ben-Dov, I.Z., Whalen, V.M., Goilav, B., Max, K.E., Tuschl, T.: Cell and microvesicle urine microRNA deep sequencing profiles from healthy individuals: observations with potential impact on biomarker studies. PLoS One. **11**(1), e0147249 (2016)
32. Sanz-Rubio, D., Martin-Burriel, I., Gil, A., Cubero, P., Forner, M., Khalyfa, A., et al.: Stability of circulating Exosomal miRNAs in healthy subjects. Sci. Rep. **8**(1), 10306 (2018)
33. Oeyen, E., Willems, H., Kindt, R., Sandra, K., Boonen, K., Hoekx, L., et al.: Determination of variability due to biological and technical variation in urinary extracellular vesicles as a crucial step in biomarker discovery studies. J. Extracell. Vesicles. **8**(1), 1676035 (2019)
34. Noren Hooten, N., McFarland, M.H., Freeman, D.W., Mode, N.A., Ezike, N., Zonderman, A.B., et al.: Association of extracellular vesicle protein cargo with race and clinical markers of mortality. Sci. Rep. **9**(1), 17582 (2019)
35. Gustafson, C.M., Shepherd, A.J., Miller, V.M., Jayachandran, M.: Age- and sex-specific differences in blood-borne microvesicles from apparently healthy humans. Biol. Sex Differ. **6**, 10 (2015)
36. Barreiro, K., Huber, T.B., Holthofer, H.: Isolating urinary extracellular vesicles as biomarkers for diabetic disease. Meth. Mol. Biol. (Clifton, NJ). **2067**, 175–188 (2020)
37. Zebrowska, A., Skowronek, A., Wojakowska, A., Widlak, P., Pietrowska, M.: Metabolome of exosomes: focus on vesicles released by cancer cells and present in human body fluids. Int. J. Mol. Sci. **20**(14) (2019)
38. Williams, C., Palviainen, M., Reichardt, N.C., Siljander, P.R., Falcon-Perez, J.M.: Metabolomics applied to the study of extracellular vesicles. Metabolites. **9**(11) (2019)
39. Erozenci, L.A., Bottger, F., Bijnsdorp, I.V., Jimenez, C.R.: Urinary exosomal proteins as (pan-)cancer biomarkers: insights from the proteome. FEBS Lett. **593**(13), 1580–1597 (2019)
40. Turchinovich, A., Drapkina, O., Tonevitsky, A.: Transcriptome of extracellular vesicles: state-of-the-art. Front. Immunol. **10**, 202 (2019)
41. Gezsi, A., Kovacs, A., Visnovitz, T., Buzas, E.I.: Systems biology approaches to investigating the roles of extracellular vesicles in human diseases. Exp. Mol. Med. **51**(3), 1–11 (2019)
42. Ramirez, M.I., Amorim, M.G., Gadelha, C., Milic, I., Welsh, J.A., Freitas, V.M., et al.: Technical challenges of working with extracellular vesicles. Nanoscale. **10**(3), 881–906 (2018)
43. Srinivasan, S., Yeri, A., Cheah, P.S., Chung, A., Danielson, K., De Hoff, P., et al.: Small RNA sequencing across diverse biofluids identifies optimal methods for exRNA isolation. Cell. **177**(2), 446-62. e16 (2019)
44. Mussack, V., Wittmann, G., Pfaffl, M.W.: Comparing small urinary extracellular vesicle purification methods with a view to RNA sequencing-enabling robust and non-invasive biomarker research. Biomol. Detect. Quant. **17**, 100089 (2019)
45. Cheng, L., Sun, X., Scicluna, B.J., Coleman, B.M., Hill, A.F.: Characterization and deep sequencing analysis of exosomal and non-exosomal miRNA in human urine. Kidney Int. **86**(2), 433–444 (2014)
46. Srinivasan, S., Duval, M.X., Kaimal, V., Cuff, C., Clarke, S.H.: Assessment of methods for serum extracellular vesicle small RNA sequencing to support biomarker development. J. Extracell. Vesicles. **8**(1), 1684425 (2019)
47. Gunasekaran, P.M., Luther, J.M., Byrd, J.B.: For what factors should we normalize urinary extracellular mRNA biomarkers? Biomol. Detect. Quant. **17**, 100090 (2019)
48. Git, A., Dvinge, H., Salmon-Divon, M., Osborne, M., Kutter, C., Hadfield, J., et al.: Systematic comparison of microarray profiling, real-time PCR, and next-generation sequencing technologies for measuring differential microRNA expression. RNA (New York, NY). **16**(5), 991–1006 (2010)
49. Hill, A.F., Pegtel, D.M., Lambertz, U., Leonardi, T., O'Driscoll, L., Pluchino, S., et al.: ISEV position paper: extracellular vesicle RNA analysis and bioinformatics. J. Extracell. Vesicles. **2** (2013)
50. Mateescu, B., Kowal, E.J., van Balkom, B.W., Bartel, S., Bhattacharyya, S.N., Buzas, E.I., et al.: Obstacles and opportunities in the functional analysis of extracellular vesicle RNA – An ISEV position paper. J. Extracell. Vesicles. **6**(1), 1286095 (2017)
51. Xu, X., Barreiro, K., Musante, L., Kretz, O., Lin, H., Zou, H., et al.: Management of Tamm-Horsfall protein for reliable urinary analytics. Proteomics Clin. Appl. **13**(6), e1900018 (2019)
52. Tian, Y., Gong, M., Hu, Y., Liu, H., Zhang, W., Zhang, M., et al.: Quality and efficiency assessment of six extracellular vesicle isolation methods by nano-flow cytometry. J. Extracell. Vesicles. **9**(1), 1697028 (2020)
53. Patel, G.K., Khan, M.A., Zubair, H., Srivastava, S.K., Khushman, M., Singh, S., et al.: Comparative analysis of exosome isolation methods using culture supernatant for optimum yield, purity and downstream applications. Sci. Rep. **9**(1), 5335 (2019)

54. Sunkara, V., Kim, C.J., Park, J., Woo, H.K., Kim, D., Ha, H.K., et al.: Fully automated, label-free isolation of extracellular vesicles from whole blood for cancer diagnosis and monitoring. Theranostics. **9**(7), 1851–1863 (2019)
55. Chiriaco, M.S., Bianco, M., Nigro, A., Primiceri, E., Ferrara, F., Romano, A., et al.: Lab-on-chip for exosomes and microvesicles detection and characterization. Sensors (Basel, Switzerland). **18**(10) (2018)
56. Ramshani, Z., Zhang, C., Richards, K., Chen, L., Xu, G., Stiles, B.L., et al.: Extracellular vesicle microRNA quantification from plasma using an integrated microfluidic device. Comm. Biol. **2**, 189 (2019)
57. Wang, S., Khan, A., Huang, R., Ye, S., Di, K., Xiong, T., et al.: Recent advances in single extracellular vesicle detection methods. Biosens. Bioelectron. **154**, 112056 (2020)
58. Babicki, S., Arndt, D., Marcu, A., Liang, Y., Grant, J.R., Maciejewski, A., et al.: Heatmapper: web-enabled heat mapping for all. Nucleic Acids Res. **44**(W1), W147–W153 (2016)
59. Zhou, G., Xia, J.: Using OmicsNet for network integration and 3D visualization. Curr. Protoc. Bioinformatics. **65**(1), e69 (2019)
60. Ghai, V., Wu, X., Bheda-Malge, A., Argyropoulos, C.P., Bernardo, J.F., Orchard, T., et al.: Genome-wide profiling of urinary extracellular vesicle microRNAs associated with diabetic nephropathy in type 1 diabetes. Kidney Int. Rep. **3**(3), 555–572 (2018)
61. Khurana, R., Ranches, G., Schafferer, S., Lukasser, M., Rudnicki, M., Mayer, G., et al.: Identification of urinary exosomal noncoding RNAs as novel biomarkers in chronic kidney disease. RNA (New York, NY). **23**(2), 142–152 (2017)
62. Salih, M., Demmers, J.A., Bezstarosti, K., Leonhard, W.N., Losekoot, M., van Kooten, C., et al.: Proteomics of urinary vesicles links plakins and complement to polycystic kidney disease. J Am Soc Nephrol. **27**(10), 3079–3092 (2016)
63. Puhka, M., Takatalo, M., Nordberg, M.E., Valkonen, S., Nandania, J., Aatonen, M., et al.: Metabolomic profiling of extracellular vesicles and alternative normalization methods reveal enriched metabolites and strategies to study prostate cancer-related changes. Theranostics. **7**(16), 3824–3841 (2017)
64. Clos-Garcia, M., Loizaga-Iriarte, A., Zuniga-Garcia, P., Sanchez-Mosquera, P., Rosa Cortazar, A., Gonzalez, E., et al.: Metabolic alterations in urine extracellular vesicles are associated to prostate cancer pathogenesis and progression. J. Extracell. Vesicles. **7**(1), 1470442 (2018)
65. Skotland, T., Ekroos, K., Kauhanen, D., Simolin, H., Seierstad, T., Berge, V., et al.: Molecular lipid species in urinary exosomes as potential prostate cancer biomarkers. Eur. J. Cancer (Oxford, England: 1990). **70**, 122–132 (2017)

Cardiac Troponin T: The Impact of Posttranslational Modifications on Analytical Immunoreactivity in Blood up to the Excretion in Urine

4

Douwe de Boer, Alexander S. Streng, William P. T. M. van Doorn, Wim H. M. Vroemen, Otto Bekers, Will K. W. H. Wodzig, and Alma M. A. Mingels

Abstract

Cardiac troponin T (cTnT) is a sensitive and specific biomarker for detecting cardiac muscle injury. Its concentration in blood can be significantly elevated outside the normal reference range under several pathophysiological conditions. The classical analytical method in routine clinical analysis to detect cTnT in serum or plasma is a single commercial immunoassay, which is designed to quantify the intact cTnT molecule. The targeted epitopes are located in the central region of the cTnT molecule. However, in blood cTnT exists in different biomolecular complexes and proteoforms: bound (to cardiac troponin subunits or to immunoglobulins) or unbound (as intact protein or as proteolytic proteoforms). While proteolysis is a principal posttranslational modification (PTM), other confirmed PTMs of the proteoforms include *N*-terminal initiator methionine removal, *N*-acetylation, *O*-phosphorylation, *O*-(*N*-acetyl)-glucosaminylation, *N(ε)*-(carboxymethyl) lysine modification and citrullination. The immunoassay probably detects several of those cTnT biomolecular complexes and proteoforms, as long as they have the centrally targeted epitopes in common. While analytical cTnT immunoreactivity has been studied predominantly in blood, it can also be detected in urine, although it is unclear in which proteoform cTnT immunoreactivity is present in

D. de Boer (✉) · W. K. W. H. Wodzig
Unit of Clinical Chemistry, Central Diagnostic Laboratory, Maastricht University Medical Center, Maastricht, The Netherlands

NUTRIM, School of Nutrition and Translational Research in Metabolism, Medicine and Life Sciences (FHML) of Maastricht University (UM), Maastricht University Medical Center, Maastricht, The Netherlands
e-mail: douwe.de.boer@mumc.nl

A. S. Streng · W. P. T. M. van Doorn
W. H. M. Vroemen
Unit of Clinical Chemistry, Central Diagnostic Laboratory, Maastricht University Medical Center, Maastricht, The Netherlands

O. Bekers · A. M. A. Mingels
Unit of Clinical Chemistry, Central Diagnostic Laboratory, Maastricht University Medical Center, Maastricht, The Netherlands

CARIM, School for Cardiovascular Diseases, Faculty of Health, Medicine and Life Sciences (FHML) of Maastricht University (UM), Maastricht University Medical Center, Maastricht, The Netherlands

H. M. Baptista Carreira dos Santos (ed.), *Translational Urinomics*, Advances in Experimental Medicine and Biology 1306, https://doi.org/10.1007/978-3-030-63908-2_4

urine. This review presents an overview of the current knowledge on the pathophysiological lifecycle of cTnT. It provides insight into the impact of PTMs, not only on the analytical immunoreactivity, but also on the excretion of cTnT in urine as one of the waste routes in that lifecycle. Accordingly, and after isolating the proteoforms from urine of patients suffering from proteinuria and acute myocardial infarction, the structures of some possible cTnT proteoforms are reconstructed using mass spectrometry and presented.

Keywords

Cardiac troponin T · Post-translational modifications · Immunoreactivity: blood · Urine

4.1 Basic Primary Structure and Isoforms of cTnT

The troponin complex, which is composed of troponin T (TnT), troponin I (TnI) and troponin C (TnC) is involved in muscle contraction. Immunoassays of the cardiac-specific variants of TnT and TnI (cTnT and cTnI) play an important role in the diagnosis of acute coronary syndrome, predominantly myocardial infarction (MI). While several reviews in respect to cTnT already have addressed all kinds of analytical, biochemical and clinical aspects in blood [1–3], this overview intends to present an integrated biochemical concept for the pathophysiological lifecycle of cTnT with a focus on posttranslational modifications (PTMs), analytical immunoreactivity and urine as one of the waste routes in that lifecycle.

In contrast to cTnI, the concentration of cTnT in serum or plasma can be measured by only one particular commercial immunoassay (Roche Diagnostics). This immunoassay is a sandwich-type assay and is targeting epitopes located in the central region of the cTnT molecule (Fig. 4.1). It is designed to quantify the intact cTnT molecule, as it is calibrated using intact recombinant cTnT [4]. These epitopes are the same for all versions of the low- as well as high-sensitivity cTnT Roche Diagnostics immunoassay. Consequently, the analytical specifity did not change between the versions. However, this cTnT assay also detects degraded cTnT forms, as long as it has the required central located epitopes. Degradation of cTnT is now a well-established phenomenon and consequently, detection of degraded cTnT forms is and should be a matter of interest, though its clinical effect remains still to be investigated. Moreover, any other possible existing PTM of the cTnT proteoforms within the region of the epitopes may be relevant for the detection by the immunoassay. Proteoforms are defined in this context as any specific molecular form of a protein product arising from a single gene and differing due to genetic variation, alternatively spliced ribonucleic acid (RNA) transcripts and PTMs [5].

The current version of the high-sensitivity cTnT (hs-cTnT) immunoassay detects cTnT immunoreactivity at the level of 87 fmol/L (3 ng/L of intact cTnT immunoreactivity equivalents when calibrated using recombinant cTnT) [6].[1] Applying this hs-cTnT immunoassay, it has been shown that the concentration of cTnT is elevated above the normal 95% reference range for healthy individuals, not only under pathological conditions, but also under some physiological conditions [4, 10]. Therefore, the clinical specificity of the hs-cTnT immunoassay is frequently subjected to discussion. It might be that under the different pathophysiological conditions or under acute and non-acute pathological circumstances the profile of biomolecular complexes and proteoforms of cTnT is distinctive and that the discussion could be settled by measuring the

[1]Although the unit mole per litre is recommended for concentrations of substances in clinical chemistry [7] and despite the fact that immunoassays are measuring purely the number of target molecules, cTnT concentrations in serum or plasma commonly are presented in ng/L [8]. Moreover, while the hs-cTnT assay is calibrated against intact recombinant human cTnT produced in *Escherichia coli* cell culture [9], this cTnT proteoform for assay calibration is not identical to the mixture of cTnT proteoforms usually measured in serum or plasma (see this review). Consequently, a cTnT concentration expressed in fmol/L is more logical than in ng/L.

```
       1-52: Highly acidic N-terminus          24-33: Peptide spliced in the isoform 6
  1 MSDIEEVVEE YEEEEQEEAA VEEEEDWRED EDEQEEAAEE DAEAEAETEE TRAEEDEEEE  60

 61 EAKEAEDGPM EESKPKPRSF MPNLVPPKIP DGERVDFDDI HRKRMEKDLN ELQALIEAHF 120
                         135-141, 146-157: Roche M7 and 11.7 antibodies, respectively
121 ENRKKEEEEL VSLKDRIERR RAERAEQQRI RNEREKERQN RLAEERARRE EEENRRKAED 180

181 EARKKKALSN MMHFGGYIQK QAQTERKSGK RQTEREKKKK ILAERRKVLA IDHLNEDQLR 240

241 EKAKELWQSI YNLEAEKFDL QEKFKQQKYE INVLRNRIND NQKVSKTRGK AKVTGRWK   298
```

Fig. 4.1 Basic primary structure of cardiac troponin T (cTnT): annotated amino acid sequence of the canonical isoform of cTnT. The orange highlighted area is spliced in the human adult cTnT isoform 6, while the canonical isoform corresponds to the isoform 1 (see also Table 4.1). The purple highlighted areas indicate the target epitopes of the clinical cTnT immunoassay by Roche Diagnostics

different forms of cTnT simultaneously. Especially, the proteolytic forms might be of interest [6, 11]. Hence, the analytical distinction between intact and degraded cTnT proteoforms might improve the clinical specificity of the hs-cTnT immunoassay. While this topic of clinical specificity and sensitivity is currently focused on measuring cTnT immunoreactivity in blood, it certainly will also be a topic when measuring cTnT immunoreactivity in urine becomes feasible. Besides proteolysis, other PTMs are of interest, especially since the knowledge of PTMs is expanding and the topic of cTnT and those other PTMs are underexposed. In relation to the measurement of cTnT immunoreactivity and besides the relevance of the basic primary structure and the impact of possible PTMs of cTnT, also those of the different isoforms and secundary, tertiary and quaternary structures of cTnT need attention. This chapter presents an overview of the current knowledge of the issue of immunoreactivity for the different structures of cTnT in blood and urine.

4.2 Isoforms of cTnT and Cross Reactivity

While the gene corresponding to cTnT determines its basic primary structure, different primary structures of cTnT are formed from alternative splicings. The gene itself is part of a troponin gene family and therefore, the cTnT variants also have relations with other troponin T proteins. Together these variants form the group of specific cTnT isoforms as well as the so-called subunit troponin isoforms. Three genes encode for the different subunit troponin T isoforms. Besides for cardiac isoforms (*TNNT2*), genes exist for slow skeletal muscle (*TNNT1*) and fast skeletal muscle (*TNNT3*) isoforms of TnT. The *TNNT2* gene encodes an amino acid sequence resulting in the straightforward and "unspliced" isoform of cTnT (isoform 1 or TNT1: Fig. 4.1) [2]. Alternative RNA splicing generates multiple other specific cTnT isoforms. The predominate specific isoform in normal adult heart is cTnT isoform 6, which misses the amino acids between position 24 and 33 when compared to the canonical sequence (Fig. 4.1). This non-expressed amino acid sequence is very negatively charged. Subsequently, this leads, besides to the shortening of the protein, also to a less acidic isoform 6 (Table 4.1). Physiologically, the non-expressed sequence results in an increase of myofilament sensitivity to calcium. This is essential for example in the fetal human heart, where the isoform 1 is expressed at a high level. In the immature myocardium this is functionally important, as the peak cytosolic calcium concentration transient is significantly less than that of isoform 6 [12].

Several mutations in the *TNNT2* gene are known, which might result in slightly different genetic variants of isoform 6. Some of these

Table 4.1 Characteristics of possible human cTnT isoforms and proteoforms based on the current knowledge as well as physical and physiological characteristics other relevant proteins

cTnT isoforms and proteoforms (labelled 1 to 19), human and bovine cTnI (labelled 20 to 23) and other relevant proteins (labelled A to F)	MW roughly estimated (Da)[a,b]	MW exactly averaged (Da)[c]	pI[c]	Radius (nm)[a]	Relative filtration[a]
1. Non-modified cTnT (isoform 1)	–	35,924	4.94	–	–
2. Non-modified cTnT (isoform 6)	37–40 kDa	34,590	5.13	3.4[d]	–
3. -- minus Met1	37–40 kDa	34,459	5.13	–	–
4. -- N-acetylated Ser2	37–40 kDa	34,501	<5.13	–	–
5. Non-modified cTnT (isoform 6) minus Lys298	37–40 kDa	34,462	5.09	–	–
6. -- minus Met1 and Lys298	37–40 kDa	34,331	5.09	–	–
7. -- -------------- *N*-acetylated Ser2	37–40 kDa	34,373	<5.09	–	–
8. -- minus Met1 and Lys290-Lys298	–	33,404	4.99	–	–
9. -- -------------- *N*-acetylated Ser2	–	33,446	<4.99	–	–
10. Non-modified cTnT (isoform 6) minus *N*-terminus[caspase-3]	29 kDa	26,773	9.86	–	–
11. -- minus *N*-terminus[caspase-3] and Lys298	29 kDa	26,645	9.82	–	–
12. -- minus *N*-terminus[caspase-3] and Lys290-Lys298	–	25,718	9.69	–	–
13. Non-modified cTnT (isoform 6) minus *N*-terminus[calpain-1]	–	24,518	9.99	–	–
14. -- minus *N*- terminus[calpain-1] and Lys298	–	24,390	9.96	–	–
15. -- minus *N*- terminus[calpain-1] and Lys290-Lys298	–	23,463	9.84	–	–
16. Non-modified cTnT (isoform 6) minus *C*-terminus	–	22,613	4.55	–	–
17. -- minus Met1 and *C*-terminus	–	22,482	4.55	–	–
18. -- minus *N*-terminus[caspase-3] and *C*-terminus	15–20 kDa	14,797	7.99	–	–
19. -- minus *N*-terminus[calpain-1] and *C*-terminus	15–20 kDa	12,541	9.34	–	–
20. Non-modified cTnI (human)	–	24,008	9.87	–	–
21. Non-modified cTnI minus Met1 (human)	–	23,876	9.87	–	–
22. Non-modified cTnI (bovine)	–	24,054	9.87	–	–
23. Non-modified cTnI minus Met1 (bovine)	–	23,922	9.87	2.9[e]	–
A. Non-modified albumin	69,000	66,472	5.67	3.5[f]	< 0.01
B. Non-modified hemoglobin ($\alpha_2\beta_2$-heme B)	68,000	62,602	7.52	3.2[g]	0.03
C. Non-modified ovalbumin (chicken)	43,500	42,881	5.19	3.0 ± 0.7[h]	0.22
D. Non-modified β-lactoglobulin dimer (bovine)	36,000	36,562	4.83	2.5 ± 6833[i]	0.4
E. Non-modified myoglobin	16,900	17,052	7.29	1.7 ± 0.1[j]	0.75
F. Non-modified lysozyme C	14,600	14,701	9.28	1.6 ± 0.3[k]	0.8

[a]The radius is the effective molecular radius, which is the Einstein-Stokes radius; it is the radius of a sphere that diffuses at the same rate as the substance indicated; if more than one dadius was reported, a mean radius was calculated

[b]Data in Da unless indicated otherwise

[c]Data as calculated using the compute pI/Mw tool using UniProt Knowledgebase (Swiss-Prot or TrEMBL) [The UniProt Consortium. Nucleic Acids Res 2019; 47(D1): D506–D515]

[d]Clin Chem Lab Med 2006; 44: 1422–1427

[e]J Biol Chem 1983; 258: 2951–2954

[f]Biophys J 2004; 87: 4259–4270

[g]Biophys J 2006; 91: 3014–3021

[h]Nature 1990; 347: 99–102

[i]Biophys J 2009; 97: 590–598/J Biol Chem 1966; 241: 2496–2501

[j]Biophys J 2000; 78; 719–730/J Phys Chem B 1998; 102: 10615–10621/Nat Commun 2015; 6; 6772–6779

[k]Arch Biochem Biophys 1982; 219: 89–100/Biophys J 2000; 78; 719–730/Biophys J 2009; 97: 590–598

genetic modifications have been associated with cardiomyopathies [2, 13]. Because the majority of these mutations are located in the middle region of cTnT, single amino acid substitutions and deletions in the target epitopes of the clinical cTnT immunoassay may directly affect the ability to detect cTnT by this immunoassay (Fig. 4.1). However, as no such examples have been studied and reported until now, this seems an unrelevant complication when measuring immunoreactivity in blood or urine.

On the other hand, cross reactivity between cardiac and skeletal troponins is an issue requiring more attention. For example, at high cTnT concentrations (e.g. after an acute MI) and high human skeletal muscle TnT (skTnT) concentrations (e.g. after an acute MI combined with a surgical intervention), extrarenal clearance will not dominate anymore [14]. Consequently, in urine high troponin concentrations may be expected and the question arises if cross-reactivity affects the measurement of cTnT. For the current fifth generation hs-cTnT immunoassay, the manufacturer claims that the assay does not show significant cross-reactions with other relevant human troponin isoforms; <0.07% with skTnT and <0.006% with human skeletal muscle TnI or human TnC, while with human cTnI <0.02% [15]. However, Schmid et al. described elevated cTnT concentrations in serum of patients with skeletal myopathies, in which they suggested cross-reaction with skTnT isoforms in the hs-cTnT immunoassay as the most likely cause [16]. This disagrees with Wens et al., who detected cTnT mRNA expression in skeletal muscle in patients with Pompe disease and identified one cTnT peptide by mass spectrometry in skeletal muscle tissue [17]. Also in myotonic dystrophy patients the concentration of cTnT in serum was significantly higher [18]. Moreover, using western blotting combined with the original monoclonal cTnT antibodies from the commercial hs-cTnT immunoassay (M7 antibody for capture and M11.7 antibody for detection, Roche Diagnostics), it was shown that in purified human skTnT at high skTnT levels not only skTnT was present, but also cTnT [19]. All these data suggest that not only cross-reactivity with skTnT, but also cTnT expression in skeletal muscle may contribute to elevated cTnT concentrations in patients with skeletal myopathies [20]. It thus may complicate an interpretation of a high cTnT concentration in blood as well as urine and requires full insight into the patients' medical record of possible muscle myopathic diseases. It may even require the measurement of skTnT.

4.3 Structures of cTnT beyond Its Basic Primary Structure

Structural characteristics of cTnT beyond its basic primary protein structure are of interest, especially if pathophysiological mechanisms result in characteristic structures and if PTMs are disease specific. While mutations in the *TNNT2* gene may have an effect on the secondary and tertiary structure of cTnT, no studies are available, which evaluated the influence of those structures on the performance of cTnT immunoassays. Therefore, this overview will focus on quaternary structures and PTMs involving cTnT.

4.4 Quaternary Structures: Biomolecular Complexes of cTnT

cTnT is the tropomyosin-binding subunit of the troponin complex specific to the cardiac muscle, which also binds to the subunits cTnI and non-cardiac specific TnC [1]. As a protein, cTnT is a non-complicated, medium-sized protein with a minor glycosylation profile. In the first hours after the onset of symptoms of a MI, a mixture of the ternary cardiac troponin complex (cTnT-cTnI-TnC) and the binary complex cTnI-TnC was found in the patients' plasma [21]. During the subsequent hours, the ratio ternary/binary complex progressively shifted towards the binary complex, meaning that cTnT is the first troponin subunit to dissociate from the ternary complex (Fig. 4.2). The presence of the ternary and binary complex in blood was also previously observed, but not rapidly after the onset of symptoms [22, 23]. This rapid dissociation was recently confirmed from a totally different perspective in a multi-site coronary vein sampling study [24]. In this study coronary venous system samples demonstrated that the ternary cardiac troponin complex was present in higher concentrations than in simultaneously obtained peripheral samples. While that of the ternary complex proportionally decreased due to dilution in the blood volume during its passage from the coronary to the periphery, it also disappeared over time. Moreover, the ternary complex was more prominent in an early than in a late presenting MI patient.

Both Bates et al. and Vylegzhanina et al. speculated whether the ternary complex itself could be subjected to proteolysis and for example could lose cTnI and cTnT fragments, but no evidence-based data are available [21, 23]. However, such a phenomenon is not illogical and should be anticipated (Fig. 4.2). Anticoagulants do have a direct effect on the stability of the ternary complex. Wu et al. showed that ethylenediaminetetraacetic acid (EDTA) destabilizes the complex into separate subunits, whereas heparin does not show such an effect [22]. This phenomenon relates to the function of calcium, which stabilizes the complex.

With respect to the detection of the ternary complex Vylegzhanina et al. applied *in-house* developed sandwich immunofluorescent assays [21, 25]. Others used different generations of the commercial hs-cTnT immunoassay by Roche Diagnostics [22–24]. Thus, it can be concluded that the commercial hs-cTnT immunoassay is able to detect the ternary troponin complex. No binary complexes with cTnT have been reported and therefore, it is unknown whether the hs-cTnT immunoassay is able to detect in blood a binary troponin complex with cTnT.

Additional and relevant quaternary structures are complexes, which include autoantibodies against cardiac troponins (Fig. 4.2). In blood of healthy subjects generally up to 10% of the subjects are showing antibodies against cTnT, while up to 20% are showing antibodies against cTnI [26]. The exact implication for the measurement of analytical immunoreactivity by the commercial hs-cTnT immunoassay is still unclear, but incidental cases have reported to detect complexes with autoantibodies this way [27, 28]. Also the *in-house* developed cTnT sandwich immunofluorescent assays were able to detect similar complexes with autoantibodies [25].

The possible cause of favoured clinical immunoreactivity against cTnI rather than against cTnT might be an immuno-dominant epitope within cTnI [29] or the localization of cTnI within the troponin complex [30]. Perhaps because the complexes of troponins commonly are not circulating in the extracellular compartment, the immune system might consider them as foreign invaders and as a reaction anticardiac troponin autoantibodies might be formed (Fig. 4.2).

Another cause or a co-factor might be citrullination. Citrullination, the irreversible PTM involving the conversion of arginine to citrulline

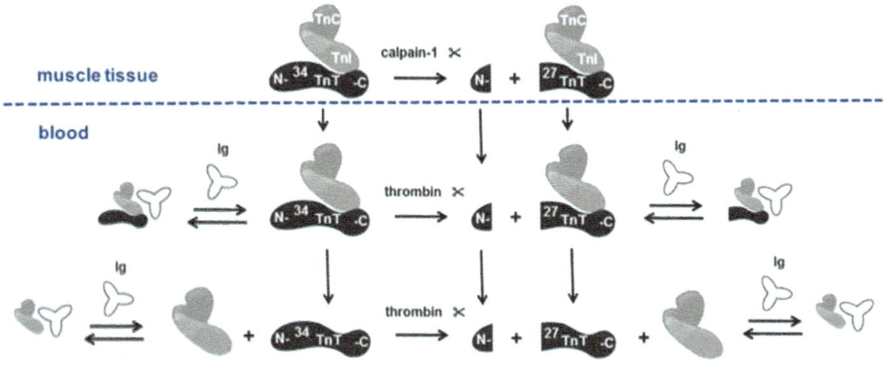

Fig. 4.2 Hypothetical model of the biomolecular complexes between cardiac troponin T (TnT) proteoforms, some proteases and autoantibody formation based on their reported presence in cardiac muscle tissue and blood; *Ig* immunoglobulin, *TnC* troponin C, *TnI* troponin I, *N-* *N*-terminus, *–C*, *C*-terminus

by the family of peptidylarginine deiminase enzymes, is associated with several diseases. Specific citrullinated proteins have been shown to alter their function while others act as autoantigens. It has been reported that citrullination of cTnI is more extensive than of cTnT, which also may contribute to a favoured clinical immunoreactivity against cTnI [31]. Within that same study Arg78 was identified in cTnTn as one of the sites that were altered in heart failure patients compared to non-failing controls.

Whether or not combined with autoantibodies, the excretion of ternary and binary troponin complexes in urine is very unlikely in patients with a normal kidney function. When the glomerular filtration has been reduced significantly, renal excretion of large protein complexes may no longer be excluded.

4.5 Posttranslational Modifications in the Basic Primary Structure of cTnT and Clearance of cTnT

PTMs in proteins include a wide variety of modifications: internal protein changes (e.g. disulfide bond formation), removal of amino acid residues (e.g. proteolysis and *N*-terminal initiator methionine removal), attachment of other proteins (e.g. ubiquitinylation, NEDDylation and SUMOylation), *N*-terminal modifications (e.g. acetylation, arginylation, lipidation by myristoylation and palmitoylation), *S*-modifications (e.g. *S*-myristoylation, *S*-sulfenylation, *S*-glutathionylation and *S*-nitrosylation) and other side-chain modifications (e.g. acylation, hydroxylation, (de)alkylation, (de)amination, oxidation, reduction, phosphorylation, glycosylation, critullination, carbamylation, lipidations such as palmitoylation and prenylation, as well as advanced glycation end product (AGE) attachment). Some of these PTMs are reversible, while others are irreversible.

PTMs in the myocardium occur in case of ischemia and cardiomyopathies on the tissue level. However, it is not always clear what the consequences are for cTnT measurements on the peripheral blood and urinary level, at which the clinical diagnosis is based on. Established PTMs for cTnT in muscle tissues consist of *N*-terminal initiator methionine removal, *N*-acetylation, *O*-phosphorylation, *O*-(*N*-acetyl)glucosaminylation, *N(ε)*-(carboxymethyl)lysine modification and citrullination. Also intracellular and extracellular proteolytic degradations are confirmed PTMs, while myristoylation is a non-confirmed PTM for cTnT. All these established PTMs for cTnT will be discussed in this review separately in the context of the pathophysiological lifecycle of cTnT.

Once released into the bloodstream, the thus generated proteoforms will undergo further proteolysis. It will generate even smaller proteoforms. Subsequently, all proteoforms will experience a different clearance process primarily dominated by liver and renal clearance. This dual clearance system has been described for myoglobin and projected on cTnT [32]. Because extrarenal clearance of cTnT dominates in rats and is not affected by renal function [33], it has been postulated that cTnT clearance occurs mainly via unidentified scavenger receptors [32]. Scavenger receptors in general bind all kinds of proteins and direct them to intracellular degradation. PTMs may accelerate such a clearance process, as previously identified for modified lipoproteins and pathogens [34]. Moreover and if escaping those receptors, maintaining appropriate levels of proteins within a body might be the drive to prevent excretion of proteins in urine. This is achieved by (1) blocking the passage through the glomerular filtration barrier, which consists of endothelium and glomerular basement membrane (GBM), (2) recovering intact proteins, which are able to pass the GBM and (3) degrading proteins after successfully passing the GBM and to recycle amino acids in order to synthesize new proteins [35]. Blocking is accomplished by an electrokinetic model, in which the filtration of ions and water will generate a potential difference and thus an electrical field across the GBM [36]. Because the majority of proteins are negatively charged in the blood circulation, a significant number of proteins will be repelled after drifting into the direction of the GBM by convection and diffusion. The remaining number of proteins will be recovered or recycled. Recovering of intact proteins by endocytosis is followed by reabsorption through transcytosis. Recycling of amino acids occurs through endocytosis of proteins followed by degradation into amino acids within lysosomes.

Therefore, the combination of all kinds of PTMs and different clearance mechanisms will assure that the pathophysiological lifecycle of cTnT is simply part of continuously recycling proteins and amino acids. The series of PTMs can be comprised into four sequentially logical steps starting in muscle tissue and ending in urine, which will be described as such. Some miscellaneous PTMs do not directly fit into the steps, but indirectly regulate or affect those steps.

4.6 Step 1: *N*-Terminal Initiator Methionine Removal, *N*-Acetylation and *C*-Terminal Proteolysis

In cardiac muscle tissue *N*-terminal initiator methionine removal is one of the possible elements of the lifecycle of cTnT and is followed by *N*-terminal acetylation [37]. Unlike *N*-side chain acetylation, *N*-terminal acetylation seems to be irreversible. Therefore, it is a one-way gateway into the pathophysiological lifecycle of cTnT. Verified by low-resolution mass spectrometric analysis the National Institute of Standards and Technology (NIST) [38] standard for the human cardiac troponin complex has been reported to contain predominately the *N*-terminal acetylated variant of the isoform 6 at Ser2 after *N*-terminal initiator methionine removal (Met1). The theoretical isoelectric point pI of the thus modified isoform 6 without *N*-acetylation is 5.13 (Table 4.1). However, *N*-acetylation neutralizes the positive charge on the *N*-terminus, therefore shifting the pI of the major intact cTnT isoform 6 variant as well as of the minor proteolytic form to be even more acidic. Consequently, the net electric charge at physiological pH makes this slightly modified cTnT molecule a relatively acidic protein like albumin. Its molecular weight (MW) with *N*-acetylation is 34,501 Da (Table 4.1). Also identified in the NIST standard is a minor proteolytic form of this proteoform with the loss of the last *C*-terminal amino acid residue (Lys298) from the full-length sequence (Fig. 4.1). Responsible intracellular proteases still have not been identified. This minor proteoform with *N*-acetylation has a theoretical pI of <5.09 and a MW of 34,373 Da. Zhang et al. confirmed all these PTMs by high-resolution mass spectrometric analysis of purified human heart tissue [39]. Consequently, the *N*-acetylated variant of the isoform 6 at Ser2 after *N*-terminal initiator methionine removal can be considered as the predominant intact form of cTnT in heart tissue.

4.7 Step 2: *N*-Ubiquinylation, *N*,*S*-Myristoylation and *O*-(*N*-Acetyl)-Glucosaminylation

N-acetylation of proteins has been suggested to precede *N*-ubiquinylation as part of a normal physiological process [37]. This process is supposed to start with *N*-terminal acetylation as a degradation signal and as a major determinant for the pathophysiological lifecycle of proteins in general and thus of cTnT in particular. The signal in its turn is thought to trigger an ubiquitin ligase, which recognizes *N*-terminal acetylated proteins and marks them with ubiquitin for destruction. Together with ubiquitin proteases, ubiquitin ligases play an essential role in degradation pathways in particular and cellular regulatory processes in general [40–42]. The principle of *N*-ubiquinylation and skeletal muscle atrophy has been studied in a mouse and rat model using mass spectrometry. In the mouse model an ubiquitin-ligase ubiquitylated certain thin filament components and promoted amongst others the degradation of slow skeletal troponins under muscle atrophy and after fasting [43]. In the rat model direct *N*-ubiquinylation of slow skeletal TnT was reported under muscle atrophy after immobilization [44]. It is very likely that *N*-ubiquitinylation of cTnT in the cardiac proteasome merely is part of a macro-molecular process, rather than a specific cTnT related process, but *N*-ubiquinylation of cTnT in the cardiac muscle is under stressful conditions a logical and predictable phenomenon.

Myristoylation of human cTnT in cardiac muscle was observed by Zhang et al. [39], but they were not able to confirm that by tandem mass spectrometry and to point out at which position in cTnT myristoylation might have occurred. Myristoylation is either a co-translational reaction that occurs for example after the removal of the *N*-terminal initiator methionine residue at an *N*-terminal glycine or a posttranslational reaction in apoptotic cells [45]. During apoptosis proteins amongst others are cleaved by caspases. This cleavage exposes an *N*-terminal glycine within a cryptic myristoylation consensus sequence, which subsequently can be myristoylated. *S*-myristoylation of proteins as a side-chain modification also has been reported. As the slightly modified variant of isoform 6 of cTnT after the removal of the *N*-terminal initiator methionine residue (Met1) remains with a serine (Ser2) and not with a glycine, a co-translational reaction *N*-myristoylation is unlikely and thus *S*-myristoylation is a more likely scenario.

O-(*N*-acetyl)-glucosaminylation or *O*-linked *β*-linked *N*-acetylglucosaminylation (*O*-Gly*N*Ac) is a PTM, which interacts with *N*-ubiquinylation [46]. Mass spectrometric proof of direct *O*-Gly*N*Ac of skTnT has been reported in rat skeletal muscle [47]. Independently from the physiological function of *N*-acetylation and *O*-GlyNAc with respect to cTnT, the above-indicated considerations imply that *N*-ubiquinylation and *S*-myristoylation must be anticipated. Therefore, future identification of cTnT immunoreactivity in blood or urine by mass spectrometry should also focus on *N*-ubiquinylated and *S*-myristoylated proteoforms of cTnT.

4.8 Miscellaneous PTMs Related to Step 2: *O*-Phosphorylation and Advanced Glycation

O-phosphorylation by kinases and dephosphorylation by phosphatases are classically known biochemical processes, which reversibly regulate the function of proteins [41]. A variety of Ser/Thr kinases regulate *O*-phosphorylation, but for humans a limited number of phosphorylated sites have been identified for cTnT [48]. An important and confirmed phosphorylation target on the isoform 6 of cTnT in human cardiac muscle is at Ser2 [39]. Site-specific effects of phosphorylation on contractile properties have been reported, mostly by using transgenic animals with cTn phosphorylation mimicking charge mutations [2, 49].

Like for *N*-ubiquinylation, *O*-Gly*N*Ac is potentially another reversible PTM of cTnT [47], which presents a dynamic interplay with phosphorylation for proteins in general [46, 50] and for cTnT in particular in ischaemic mice heart failure models [51]. However, for humans no such sites in cTnT have been identified.

Advanced glycation is the irreversible attachment of reducing sugars onto the free amino groups of proteins. Its physiological roles include the identification of aged and damaged proteins [52]. While it is interesting to hypothesize whether aged proteins contain more PTMs, which may result in protein damage and aggregation, it is clear that there is a time dependent accumulation of advanced glycation end-products (AGEs). AGE labelled proteins are catabolised by cells into peptides and amino acids. While the AGE free adducts are released directly into the urine, AGE peptides are recycled and excreted in urine to a lesser extent [53]. cTnT exhibited fewer AGE modification sites than cTnI [54]. These modifications were limited to *N(ε)*-(carboxymethyl)lysine (CML)-modification and were detected on 3 cTnT amino acid residues (Lys107, Lys125 and Lys227). These AGE peptides of cTnT may modify interaction of the troponin subunits and influence phosphorylation-mediated signalling regulation of the troponin complex [54].

It has been claimed that excessive *O*-Gly*N*Ac is a major trigger of the glucotoxic events that affect heart function under chronic hyperglycaemia [55]. Consequently, it may be clear that advanced glycation has a cardiac impact and that it may be of interest to explore CML-modification of cTnT and to evaluate these AGE peptides of cTnT in for examples diabetic subjects at high risk for subclinical cardiac injury [56]. Once more, as PTMs may accelerate clearance by scavenger receptors, it might be a challenge to detect single amino acid PTMs in relation to cTnT.

4.9 Step 3: Proteolytic Degradation

Proteolytic degradation is a dominant step in the pathophysiological lifecycle of cTnT. It can be classified in minor and major modifications by either intracellular or extracellular proteases. Intracellular proteases, that are shown to have a function in the physiology of cTnT, are calpain-1 and caspase-3. The only extracellular protease, that has been identified to play a role, is thrombin. Obviously, more proteases must be involved, but those have not yet been identified in the context of the human cTnT isoform 6. Candidate proteases, which should be investigated amongst several others, are proteases like in the digestive tract. After all, it has been reported that for example rabbit skTnT is cleaved directly by chymotrypsine into several fragments including one with the *C*-terminus [57, 58]. However, and as previously indicated, proteolysis at the *C*-terminus of cTnT already was observed in heart tissue, which suggests that proteases other than thrombin should also be looked for [38, 39].

Calpain-1, or μ-calpain according to former nomenclature, is a calcium-dependent cysteine protease and acts directly as an acute mechanism to adjust muscle contractility under stress conditions [59]. The increasing calcium concentration during ischemia and reperfusion causes an activation of calpain-1. Upon activation it affects the contractile apparatus and also impairs the energy production by cleaving structural and functional proteins of myocytes and mitochondria. Additionally, it is involved in structural remodelling after MI by the generation and release of apoptosis-inducing factors [60]. Calpain-1 has a broad endopeptidase specificity [61].

Caspase-3 is a cysteine–aspartic acid protease and is as such one of the several caspases that are involved in a chain of reactions leading to apoptotic cardiomyocyte death [62]. Its specific protease activity is based on a cysteine in its active site, which cleaves a target protein only after an Asp residue with a strict requirement for an Asp residue at positions P1 and P4 and a preferred cleavage sequence of Asp-Xaa-Xaa-Asp-|- with a hydrophobic amino-acid residue at P2 and a hydrophilic amino-acid residue at P3, although Val or Ala are also accepted at this position [63]. Interestingly, caspase-3 only cleaves cTnT in its ternary complex and not in its free form [64].

After release into the bloodstream cTnT is cleaved by the most abundant coagulation protease thrombin [65, 66]. Thrombin is a serine protease that converts amongst others fibrinogen into fibrin and therefore plays a central role in the coagulation cascade. While in fibrinogen it cleaves selectively Arg-|-Gly bonds, thrombin can cleave bonds in general after Arg and Lys [67]. The consensus thrombin cleavage site for cTnT is identical to that of calpain-1 [68]. However, further proteolysis does occur [69] and for example combined with the loss of the *C*-terminus it leads to proteoforms of cTnT in the theoretical range of 12 to 22 kDa, the so-called secondary products (Table 4.1). The majority of cTnT proteolysis studies are applying sodium dodecyl sulfate polyacrylamide gel electrophoresis (SDS-PAGE) combined with immunodetection to estimate MWs of proteoforms of cTnT [24, 25, 65, 66, 68–72]. The consequence is that the MWs of intact cTnT and the proteolytic products are overestimated. As indicated in Table 4.1, those of intact cTnT are estimated to be in the range of 37–40 kDa and those of the primary products in the range of 29 kDa. Observed secondary products are more heterogeneous and estimated to be in the range of 15 to 20 kDa [24, 25, 65, 66, 68–72]. The overestimation becomes clear when the exact theoretical MWs of different possible cTnT proteoforms are calculated, which are in the range of 34 kDa for intact cTnT, of 26 kDa for the primary products and of 12 to 22 kDa for the assumed secondary products (Table 4.1). A similar overestimation occurs when applying Size Exclusion Chromatography (SEC)[2] combined with immu-

[2] Technique is also known as Gel Filtration Chromatography (GFC).

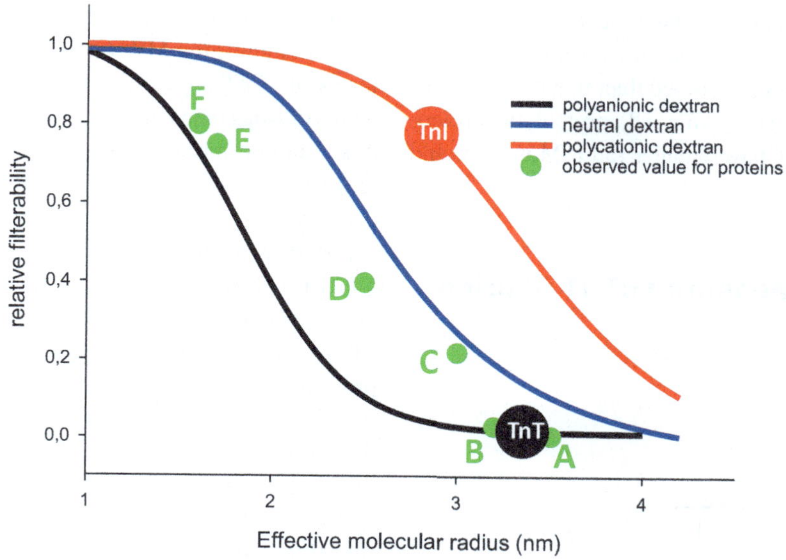

Fig. 4.3 Effect of the molecular radius and electric charge of model dextrans on the relative filterability by the glomerular capillaries as derived from data described by Hall [77]. A value of 1.0 indicates that the dextran is filtered as freely as water, whereas a value of 0 indicates that it is not filtered. Dextrans are polysaccharides that can be manufactured as neutral molecules or with negative or positive charges and with varying molecular weights. Also shown are the observed filterabilities of globular proteins (albumin [A], hemoglobin [B], ovalbumin [C], β-lactoglobulin [D], myoglobin [E], lysozyme C [F]); *TnI* cardiac troponin I, *TnT* cardiac troponin T

nodetection [22, 24, 25, 71, 73]. The elongated shape of cTnT may even lead to a misinterpretation using SEC [23, 74], because calibration of SEC commonly is based on proteins with a spherical shape [71]. Overestimation of MWs of proteoforms in biological materials can be prevented by the analysis of cTnT using mass spectrometry, but for routine analysis in for example blood or urine the analytical sensitivity of liquid tandem mass spectrometry (LC-MS/MS)-based methods is still a true challenge [68,75,76].

4.10 Step 4: Excretion in Urine

Theoretically, the excretion in urine of the intact proteoforms of cTnT by the kidney would be restricted by their relatively negative electric charge at physiological pH of intact cTnT and/or by their molecular shape and size. As its effective molecular radius characterizes the molecular shape and size of a protein objectively, this radius correlates with the renal clearance of proteins. This way, and combined with the electric charge at physiological pH, the excretion of intact cTnT and cTnI in the kidney by the glomerulus can be predicted accordingly using a model based on anionic, cationic and neutral macromolecular dextrans (Fig. 4.3).

Merely based on the MW, it implies that intact cTnT would behave as for example the bovine β-lactoglobulin dimer (Table 4.1). Because the percentage of the bovine β-lactoglobulin dimer filtered in urine is 40% and assuming that no proteolysis occurs and no other clearance mechanisms are involved, the maximum percentage of intact cTnT excreted in urine would be similar. However, and rationally, because of the elongated shape of cTnT, which results in a relatively high effective molecular radius, the percentage probably would be more likely as that of albumin and thus <1% (Fig. 4.3 and Table 4.1). This is in contrast to cTnI, which would have an excretion percentage in the range of 70%. Although the true excretion percentages may differ from the model-based percentages, it is very probable that due to its physical characteristics intact cTnT cannot be filtered in the kidney by the glomerulus.

Obviously and also theoretically, the excretion in urine of the proteolytic proteoforms of cTnT by the kidney is less restricted than that of intact proteoforms of cTnT. After all, proteolytic proteoforms of cTnT have smaller radii. Moreover, and based on proteolytic activities and thus by losing the *N*-terminus, the MW of the respective proteoforms of cTnT are in the range of 23 to 27 kDa with a drastic change in pI (Table 4.1 and Fig. 4.5). Assuming no further proteolysis the combination of a smaller size and an increased pI must result that these primary proteolytic products are excreted in a higher fraction in urine, probably >40% (Table 4.1). As indicated, further proteolysis does occur and in principal leads to proteoforms of cTnT in the theoretical range of 12–22 kDa. These products can be excreted in urine without further restrictions. Consequently, it can be summarized, that proteolysis leads to smaller proteoforms and that they easily can be excreted in urine.

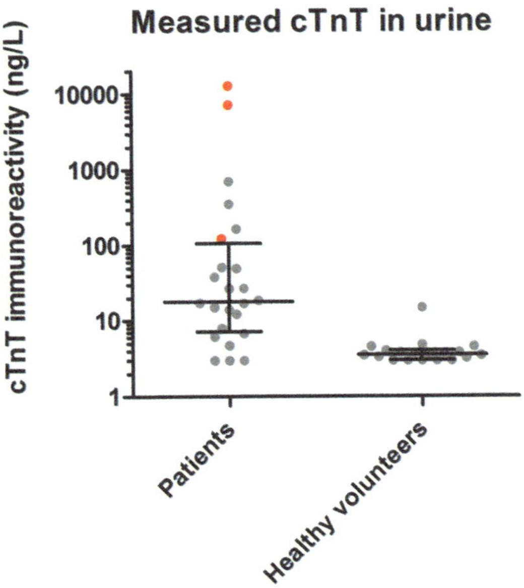

Fig. 4.4 Cardiac troponin T (cTnT) immunoreactivity as detected in urine samples of myocardial infarction patients and healthy volunteers. Red dots indicate the results of urine samples where cTnT was identified using mass spectrometry (reproduced with permission) [76]

Lastly, Van der Linden et al. from our research group argued that the contribution of renal cTnT clearance is low based on the diurnal variation of cTnT in a group of healthy subjects and chronic kidney disease (CKD) patients [78]. Using a validated mathematical model, it was predicted that the amplitude of the observed diurnal variation would fade out in patients with renal disease. Because no such fading was observed, we concluded that impaired renal elimination is not the main driver behind increased cTnT levels in patients with CKD. Based on the thoughts of the present chapter, the CKD results do not really exclude that cTnT is cleared by the kidneys,

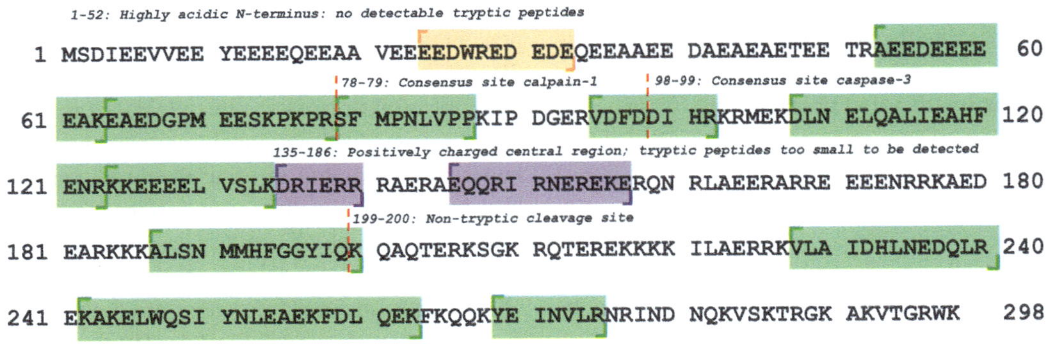

Fig. 4.5 Relation between the analytical as well as pathophysiological activity of proteases and targeted peptides in cardiac troponin T (cTnT) analysis: annotated amino acid sequence of the canonical isoform of cTnT. Green highlighted areas indicate peptides of interest that were synthesized for and targeted in a liquid tandem mass spectrometry (LC-MS/MS) method for the cTnT analysis in urine [76]. The orange highlighted area is spliced in the human adult cTnT isoform 6. The purple highlighted areas indicate the target epitopes of the clinical cTnT immunoassay by Roche Diagnostics

but merely suggests that factors other than renal elimination such as subclinical MI might contribute to elevated cTnT levels in subjects with CKD [78].

4.11 Observed Proteoforms in Urine

The concentration of cTnT in urine has been observed and evaluated in only a very few reported studies [76, 79]. In our study a targeted LC-MS/MS method was developed for the analysis of cTnT in serum, which was combined with a specific immunoprecipitation assay employing the M11.7 antibody containing the EQQRIRNEREKE epitope (Fig. 4.5) as well as with a non-specific acetone protein precipitation in order to isolate cTnT proteoforms from urine [76]. The LC-MS/MS method was optimized using 12 specific cTnT peptides [75]. Ten of those peptides were monitored, or measured after the precipitation and tryptic digestion by LC-MS/MS of urine samples from patients suffering from proteinuria and MI with an increased urinary cTnT concentration as measured by the clinical cTnT immunoassay (Fig. 4.5). The amino acid coverage of the cTnT isoform 6 was <60% and was limited amongst others because of the highly acidic *N*-terminus and the positively charged central region, which both in these regions did not yield detectable tryptic peptides. However, despite of that, in a few urine samples some specific cTnT peptides were identified (Table 4.2). Compared to urine samples from healthy volunteers and also of other patients, the cTnT immunoreactivity in those few urine samples was high to exceptionally high (Fig. 4.4).

In our study of the analytical identification of specific cTnT peptides in urine of different subjects we did not speculate on the structures of the urinary cTnT proteoforms, of which the analytically identified peptides must have originally arisen from [76]. However, this still can be done based on the separate pieces of information obtained by precipitation and immunochemical and LC-MS/MS analysis. Therefore, the hypothetical structure of the possible proteoforms can be reconstructed by the combination of those pieces of information (Table 4.2). These reconstructed structures represent the minimal possible amino acid sequence present in the urine of three subjects (A, B, and C) and as such are all proteoforms which can be excreted in urine without restrictions (Fig. 4.5).

Compared to the measured analytical immunoreactivity in the urine samples of subject A and B, the number and analytical abundances of the identified peptides were disappointing (Table 4.2). Obvious reasons are an inadequate limit of detection when applying LC-MS/MS for the detection of the peptides involved or extensive proteolysis and thus preventing the detection of the relevant peptides. This limit of detection is the range of 0.6–5 μM (1200–8100 ng/L of cTnT peptide equivalents [75]). Less obvious reasons that may have affected the measurement by the hs-cTnT assay are PTMs not accounted for. In this context the CML-modifications of Lys107, Lys125 and Lys227 in cTnT may have interfered (Fig. 4.6). While elevated levels of urinary cTnT immunoreactivity were associated with patients suffering from proteinuria [76, 79], it is also well known that AGE formation directly affects renal function [80]. A disputable reason is the uncertain impact of cross-reactivity with skTnT [19], although it is also unclear whether the subjects involved, suffered from skTnT injury due to an unidentified medical intervention, which might have initiated release of skTnT. Also citrullination in combination with cTnT autoantibodies is not an obvious explanation in spite of the fact that the involvement of autoantibodies were not ruled out and renal dysfunction caused proteinuria in the subjects investigated (Fig. 4.6).

In Fig. 4.7 we hypothetically linked the pathophysiological relations between cTnT proteoforms and involved proteases to the peptides identified in urine [76]. Besides a possible link between the 15 kDa proteoform and the peptides identified in the urine of subject A and B, also a link between the *N*-terminus fragment with those of subject C is indicated.

Table 4.2 Possible proteoforms observed in the urine of patients suffering from proteinuria and an urinary elevated cTnT analytical immunoreactivity [74]

Identified peptides by LC-MS/MS in urine samples of subjects	Detected after immuno-precipitation	Detected by clinical cTnT immunoassay	Hypothetical reconstructed proteoform with a reference to Fig. 4.5 for the context of the annotated amino acid sequence of the canonical isoform of cardiac troponin T (cTnT)
Subject A			
^{79}SFMPNLVPPK^{88} ^{95}VDFDDIHR^{102} ^{108}DLNELQALIEAHFENR^{123} ^{124}KKEEEELVSLK^{13}	YES	YES 13,000 ng/L	79 SF MPNLVPPKIP DGERVDFDDI HRKRMEKDLN ELQALIEAHF 120 121 ENRKKEEEEL VSLKDRIERR RAERAEQQRI RNEREKE 157
Subject B			
^{124}KKEEEELVSLK^{13}	YES	YES 7365 ng/L	124 KKEEEEL VSLKDRIERR RAERAEQQRI RNEREKE 157
Subject C			
^{53}AEEDEEEEAK^{63} ^{64}EAEDGPMEESKPKPR^{78}	NO	YES 125 ng/L	53 AEEDEEEE EAKEAEDGPM EESKPKPR 78

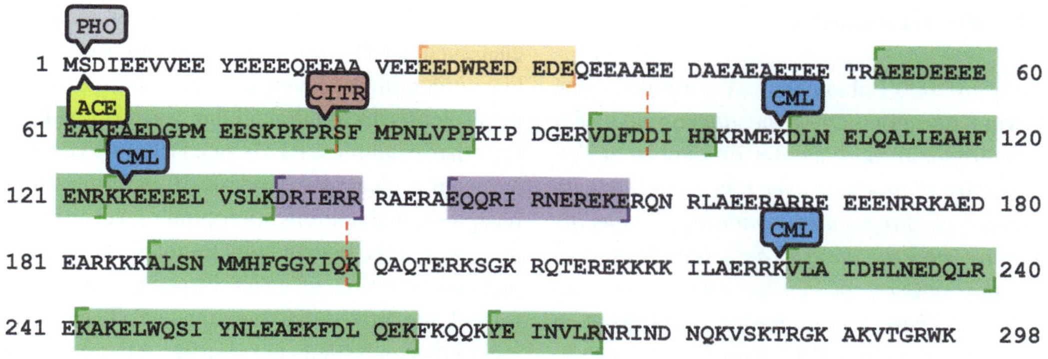

Fig. 4.6 Relation between established posttranslational modifications (PTMs) and targeted peptides in cardiac troponin T (cTnT) analysis: annotated amino acid sequence of the canonical isoform of cTnT. Green highlighted areas indicate peptides of interest that were targeted in a liquid tandem mass spectrometry (LC-MS/MS) method for the cTnT analysis in urine [76]. The orange highlighted area is spliced in the human adult cTnT isoform 6. The purple highlighted areas indicate the target epitopes of the clinical cTnT immunoassay by Roche Diagnostics. Other labelled and reported positions indicate PTMs, that have been established in literature: phosphorylation (PHO), *N*-acetylation (ACE) after *N*-terminal initiator methionine removal, citrullination (CITR) and *N(ε)*-(carboxymethyl)lysine (CML)-modification (see this review)

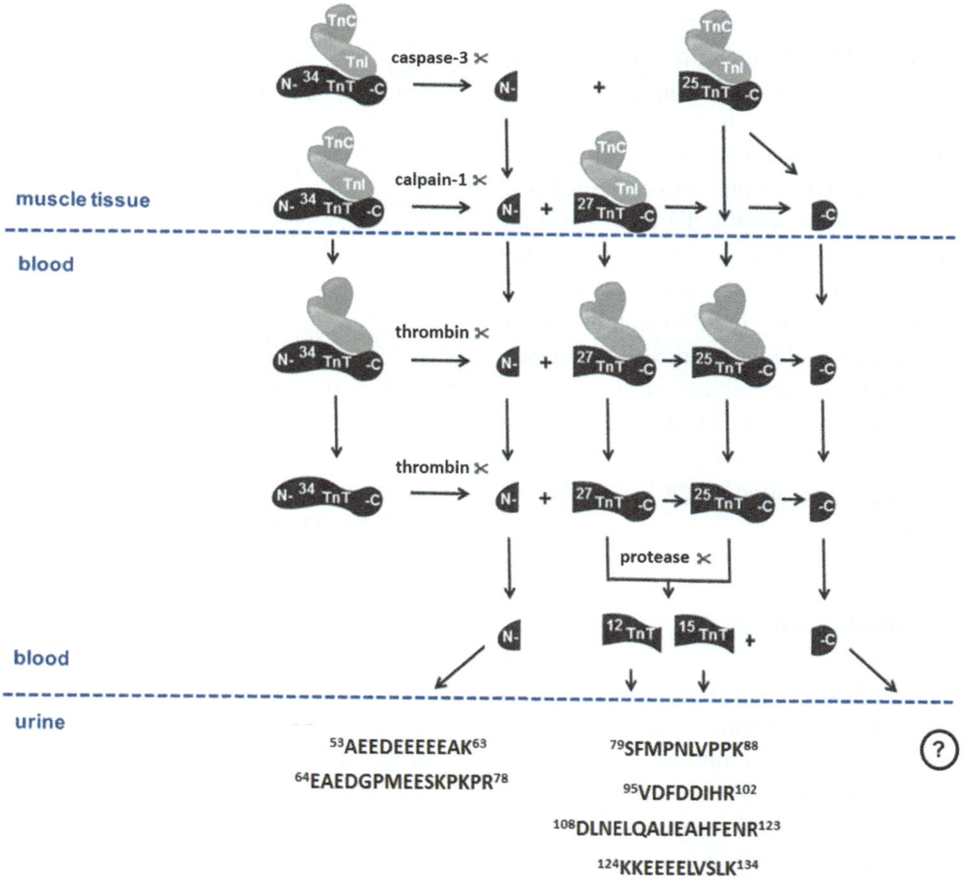

Fig. 4.7 Hypothetical and simplified model of the pathophysiological relations between cardiac Troponin T (TnT) proteoforms and involved proteases based on their reported presence in cardiac muscle tissue and blood as well as the detected peptides in urine; *TnC* troponin C, *TnI* troponin I, *N-* *N*-terminus, *–C* *C*-terminus

4.12 Final Remarks

Already in 2000 Labugger et al. formulated the following question: *Does the appearance of a certain troponin modification product or a distinct pattern of products over time correlate with a distinct cardiovascular condition, a specific time point after the onset of an acute MI or, possibly, the severity of an infarct or even re-infarction? These issues must be addressed in larger clinical studies* [81]. Anno 2020, as indicated in this overview, it is now obvious that in blood cTnT is subjected to a cascade of proteolytic cleavages. As such, important answers to the question of Labugger et al. can be given. The next thing would be to include more PTMs in investigations, especially since cTnT and PTMs other than proteolysis are an underexposed topic. Although our study was the first attempt to tackle the problem of analytical cTnT immunoreactivity in urine by LC-MS/MS, it might be that targeting unmodified peptides is an incomplete analytical strategy. Figure 4.6 summarizes the identified PTMs for cTnT as specified in this overview in relation to the targeted peptides by us [76]. It also implies that as far as known the epitopes of the clinical assay of Roche Diagnostics are not affected by those PTMs. Future LC-MS/MS investigations of cTnT proteoforms in blood and well as in urine should therefore also target peptides with PTMs in general and citrullinated and CML-modified peptides in particular. This way the problem of unexplained analytical cTnT immunoreactivity in urine might be resolved by LC/MS/MS.

4.13 Conclusion

In blood cTnT exists in different biomolecular complexes and proteoforms: bound (to cardiac troponin subunits or to immunoglobulins) or unbound (as intact protein or as proteolytic proteoforms). The hs-cTnT immunoassay probably detects several of those cTnT biomolecular complexes and proteoforms, as long as they have the central region and thus the same epitopes in common and the epitopes are not modified. All proteoforms will experience different clearance processes primarily dominated by liver and renal clearance. cTnT clearance might possibly occur via unidentified scavenger receptors and direct the proteoforms to intracellular degradation. Like for other examples, PTMs in cTnT may accelerate such a clearance process. If escaping those receptors, maintaining appropriate levels of proteins within the human body might be the drive to prevent excretion of cTnT in urine. Despite of this all, we proved that analytical cTnT immunoreactivity can be detected in urine. After isolating the proteoforms from urine of patients suffering from proteinuria and acute myocardial infarction, the structures of some possible cTnT proteoforms were reconstructed using mass spectrometry.

References

1. Katrukha, I.A.: Human cardiac troponin complex. Structure and functions. Biochemistry (Mosc). **78**, 1447–1465 (2013)
2. Wei, B., Jin, J.P.: TNNT1, TNNT2, and TNNT3: isoform genes, regulation, and structure-function relationships. Gene. **582**, 1–13 (2016)
3. Westermann, D., Neumann, J.T., Sörensen, N.A., Blankenberg, S.: High-sensitivity assays for troponin in patients with cardiac disease. Nat. Rev. Cardiol. **14**, 472–483 (2017)
4. Giannitsis, E., Kurz, K., Hallermayer, K., Jarausch, J., Jaffe, A.S., Katus, H.A.: Analytical validation of a high-sensitivity cardiac troponin T assay. Clin. Chem. **56**, 254–261 (2010)
5. Smith, L.M., Kelleher, N.L., The Consortium for Top Down Proteomics: Proteoform: a single term describing protein complexity. Nat. Methods. **10**, 186–187 (2013)
6. Van Doorn, W., Vroemen, W., De Boer, D., Mingels, A., Bekers, O., Wodzig, W., Meex, S.: Clinical laboratory practice recommendations for high-sensitivity cardiac troponin testing. J. Lab. Precis Med. **3**, 1–5 (2018)
7. IUPAC Recommendations 1979. International Union of Pure and Applied Chemistry (IUPAC) Commission on Quantities and Units in Clinical Chemistry/ International Federation of Clinical Chemistry Expert Panel on Quantities and Units. Quantities and units in clinical chemistry. Pure Appl. Chem. **1**, 2451–2479 (1979), and Clin. Chim. Acta. **96**, 157F–183F (1979)
8. Apple, F.S., Collinson, P.O., IFCC Task Force on Clinical Applications of Cardiac Biomarkers: Analytical characteristics of high-sensitivity cardiac troponin assays. Clin. Chem. **58**, 54–61 (2012)
9. Hallermayer, K., Klenner, D., Vogel, R.: Use of recombinant human cardiac troponin T for standardization of third generation troponin T methods. Scand. J. Clin. Lab. Invest. Suppl. **230**, 128–131 (1999)

10. Aakre, K.M., Omland, T.: Physical activity, exercise and cardiac troponins: clinical implications. Prog. Cardiovasc. Dis. **62**, 108–115 (2019)
11. DeFilippi, C., Seliger, S.: The cardiac troponin renal disease diagnostic conundrum: past, present, and future. Circulation. **137**, 452–454 (2018)
12. Adamcová, M., Štěrba, M., Šimůnek, T., Potáčová, A., Popelová, O., Geršl, V.: Myocardial regulatory proteins and heart failure. Eur. J. Heart Fail. **8**, 333–342 (2006)
13. Yu, B., Barbalic, M., Brautbar, A., Nambi, V., Hoogeveen, R.C., Tang, W., Mosley, T.H., Rotter, J.I., deFilippi, C.R., O'Donnell, R.J., Kathiresan, S., Rice, K., Heckbert, S.R., Ballantyne, C.M., Psaty, B.M., Boerwinkle, E.: Association of genome-wide variation with highly sensitive cardiac troponin-T (hs-cTnT) levels in European- and African-Americans: a meta-analysis from the atherosclerosis risk in communities and the cardiovascular health studies. Circ. Cardiovasc. Genet. **6**, 82–88 (2013)
14. Mair, J., Lindahl, B., Hammarsten, O., Müller, C., Giannitsis, E., Huber, K., Möckel, M., Plebani, M., Thygesen, K., Jaffe, A.S.: How is cardiac troponin released from injured myocardium? Eur. Heart J. Acute Cardiovasc. Care. **7**, 553–560 (2018)
15. Package Insert hs-cTnT assay, Roche diagnostics, version 2019-12, V 3.0 English, 2019
16. Schmid, J., Liesinger, L., Birner-Gruenberger, R., Stojakovic, T., Scharnagl, H., Dieplinger, B., Asslaber, M., Radl, R., Beer, M., Polacin, M., Mair, J., Szolar, D., Berghold, A., Quasthoff, S., Binder, J.S., Rainer, P.P.: Elevated cardiac troponin T in patients with skeletal myopathies. J. Am. Coll. Cardiol. **71**, 1540–1549 (2018)
17. Wens, S.C., Schaaf, G.J., Michels, M., Kruijshaar, M.E., Van Gestel, T.J., In 't Groen, S., Pijnenburg, J., Dekkers, D.H., Demmers, J.A., Verdijk, L.B., Brusse, E., van Schaik, R.H., Van der Ploeg, A.T., Van Doorn, P.A., Pijnappel, W.W.: Elevated plasma cardiac troponin T levels caused by skeletal muscle damage in Pompe disease. Circ. Cardiovasc. Genet. **9**, 6–13 (2016)
18. Valaperta, R., Gaeta, M., Cardani, R., Lombardi, F., Rampoldi, B., De Siena, C., Mori, F., Fossati, B., Gaia, P., Ferraro, O.E., Villani, S., Iachettini, S., Piccoli, M., Cirillo, F., Pusineri, E., Meola, G., Costa, E.: High-sensitive cardiac troponin T (hs-cTnT) assay as serum biomarker to predict cardiac risk in myotonic dystrophy: a case-control study. Clin. Chim. Acta. **463**, 122–128 (2016)
19. Vroemen, W.H.M., De Boer, D., Streng, A.S., Mingels, A.M.A., Meex, S.J.R.: Elevated cardiac troponin T in skeletal myopathies: skeletal TnT cross-reactivity and/or cardiac TnT expression? J. Am. Coll. Cardiol. **72**, 347–349 (2018)
20. Du Fay de Lavallaz, J., Zehntner, T., Puelacher, C., Walter, J., Strebel, I., Rentsch, K., Boeddinghaus, J., Nestelberger, T., Twerenbold, R., Mueller, C.: Rhabdomyolysis: a noncardiac source of increased circulating concentrations of cardiac troponin T? J. Am. Coll. Cardiol. **72**(23 Pt A), 2936–2937 (2018)
21. Vylegzhanina, A.V., Kogan, A.E., Katrukha, I.A., Koshkina, E.V., Bereznikova, A.V., Filatov, V.L., Bloshchitsyna, M.N., Bogomolova, A.P., Katrukha, A.G.: Full-size and partially truncated cardiac troponin complexes in the blood of patients with acute myocardial infarction. Clin. Chem. **65**, 882–892 (2019)
22. Wu, A.H., Feng, Y.J., Moore, R., Apple, F.S., McPherson, P.H., Buechler, K.F., Bodor, G.: Characterization of cardiac troponin subunit release into serum after acute myocardial infarction and comparison of assays for troponin T and I. American Association for Clinical Chemistry Subcommittee on cTnI standardization. Clin. Chem. **44**, 1198–1208 (1998)
23. Bates, K.J., Hall, E.M., Fahie-Wilson, M.N., Kindler, H., Bailey, C., Lythall, D., Lamb, E.J.: Circulating immunoreactive cardiac troponin forms determined by gel filtration chromatography after acute myocardial infarction. Clin. Chem. **56**, 952–958 (2010)
24. Damen, S.A.J., Vroemen, W.H.M., Brouwer, M.A., Mezger, S.T.P., Suryapranata, H., Van Royen, N., Bekers, O., Meex, S.J.R., Wodzig, W.K.W.H., Verheugt, F.W.A., De Boer, D., Cramer, G.E., Mingels, A.M.A.: Multi-site coronary vein sampling study on cardiac troponin T degradation in non-ST-segment-elevation myocardial infarction: toward a more specific cardiac troponin T assay. J. Am. Heart Assoc. **8**, e012602 (2019)
25. Vylegzhanina, A.V., Kogan, A., Katrukha, I.A., Antipova, O.V., Kara, A.N., Bereznikova, A.V., Koshkina, E.V., Katrukha, A.G.: Anti-cardiac troponin autoantibodies are specific to the conformational epitopes formed by cardiac troponin I and troponin T in the ternary troponin complex. Clin. Chem. **63**, 343–350 (2017)
26. Vilela, E.M., Bettencourt-Silva, R., Da Costa, J.T., Barbosa, A.R., Silva, M.P., Teixeira, M., Primo, J., Gama Ribeiro, V., Nunes, J.P.L.: Anti-cardiac troponin antibodies in clinical human disease: a systematic review. Ann. Transl. Med. **5**, 307–322 (2017)
27. Michielsen, E.C., Bisschops, P.G., Janssen, M.J.: False positive tropoinin result caused by a true macrotroponin. Clin. Chem. Lab. Med. **49**, 923–925 (2011)
28. Aliste-Fernández, M., Sole-Enrech, G., Cano-Corres, R., Teodoro-Marin, S., Berlanga-Escalera, E.: Description of interference in the measurement of troponin T by a high-sensitivity method. Biochem. Med. (Zagreb). **29**, 1–7 (2019)
29. Kaya, Z., Katus, H.A., Rose, N.R.: Cardiac troponins and autoimmunity: their role in the pathogenesis of myocarditis and of heart failure. Clin. Immunol. **134**, 80–88 (2010)
30. Okazaki, T., Tanaka, Y., Nishio, R., Mitsuiye, T., Mizoguchi, A., Wang, J., Ishida, M., Hiai, H., Matsumori, A., Minato, N., Honjo, T.: Autoantibodies against cardiac troponin I are responsible for dilated cardiomyopathy in PD-1-deficient mice. Nat. Med. **9**, 1477–1483 (2003)

31. Fert-Bober, J., Giles, J.T., Holewinski, R.J., Kirk, J.A., Uhrigshardt, H., Crowgey, E.L., Andrade, F., Bingham 3rd, C.O., Park, J.K., Halushka, M.K., Kass, D.A., Bathon, J.M., Van Eyk, J.E.: Citrullination of myofilament proteins in heart failure. Cardiovasc. Res. **108**, 232–242 (2015)
32. Hammarsten, O., Mair, J., Möckel, M., Lindahl, B., Jaffe, A.S.: Possible mechanisms behind cardiac troponin elevations. Biomarkers. **23**, 725–734 (2018)
33. Fridén, V., Starnberg, K., Muslimovic, A., Ricksten, S.E., Bjurman, C., Forsgard, N., Wickman, A., Hammarsten, O.: Clearance of cardiac troponin T with and without kidney function. Clin. Biochem. **50**, 468–474 (2017)
34. Zani, I.A., Stephen, S.L., Mughal, N.A., Russell, D., Homer-Vanniasinkam, S., Wheatcroft, S.B., Ponnambalam, S.: Scavenger receptor structure and function in health and disease. Cell. **4**, 178–201 (2015)
35. Saritas, T., Kuppe, C., Moeller, M.J.: Progress and controversies in unraveling the glomerular filtration mechanism. Curr. Opin. Nephrol. Hypertens. **24**, 208–216 (2015)
36. Comper, W.D., Russo, L.M., Vuchkova, J.: Are filtered plasma proteins processed in the same way by the kidney? J. Theor. Biol. **410**, 18–24 (2016)
37. Lee, K.E., Heo, J.E., Kim, J.M., Hwang, C.S.: N-terminal acetylation-targeted N-end rule proteolytic system: the Ac/N-end rule pathway. Mol. Cells. **39**, 169–178 (2016)
38. Gonzalez C, Choquette SJ. National Institute of Standards & Technology (NIST), Certificate of Analysis, Standard Reference Material® 2921, Human Cardiac Troponin Complex, Certificate Issue March 15, 2016
39. Zhang, J., Zhang, H., Ayaz-Guner, S., Chen, Y.C., Dong, X., Xu, Q., Ge, Y.: Phosphorylation, but not alternative splicing or proteolytic degradation, is conserved in human and mouse cardiac troponin T. Biochemistry. **50**, 6081–6092 (2011)
40. Chaugule, V.K., Walden, H.: Specificity and disease in the ubiquitin system. Biochem. Soc. Trans. **44**, 212–227 (2016)
41. Smoly, I., Shemesh, N., Ziv-Ukelson, M., Ben-Zvi, A., Yeger-Lotem, E.: An asymmetrically balanced organization of kinases versus phosphatases across eukaryotes determines their distinct impacts. PLoS Comput. Biol. **13**, e1005221 (2017)
42. Ebner, P., Versteeg, G.A., Ikeda, F.: Ubiquitin enzymes in the regulation of immune responses. Crit. Rev. Biochem. Mol. Biol. **52**, 425–460 (2017)
43. Cohen, S., Zhai, B., Gygi, S.P., Goldberg, A.L.: Ubiquitylation by Trim32 causes coupled loss of desmin, Z-bands, and thin filaments in muscle atrophy. J. Cell Biol. **198**, 575–589 (2012)
44. Ryder, D.J., Judge, S.M., Beharry, A.W., Farnsworth, C.L., Silva, J.C., Judge, A.R.: Identification of the acetylation and ubiquitin-modified proteome during the progression of skeletal muscle atrophy. PLoS One. **10**, e0136247 (2015)
45. Martin, D.D., Beauchamp, E., Berthiaume, L.G.: Post-translational myristoylation: fat matters in cellular life and death. Biochimie. **93**, 18–31 (2011)
46. Zachara, N., Akimoto, Y., Hart, G.W.: Chapter 19, The O-GlcNAc modification. In: Varki, A., Cummings, R.D., Esko, J.D., et al. (eds.) Essentials of Glycobiology [Internet], 3rd edn. Cold Spring Harbor Laboratory Press, Cold Spring Harbor (2015–2017)
47. Cieniewski-Bernard, C., Montel, V., Berthoin, S., Bastide, B.: Increasing O-GlcNAcylation level on organ culture of soleus modulates the calcium activation parameters of muscle fibers. PLoS One. **7**, e48218 (2012)
48. Streng, A.S., De Boer, D., Van der Velden, J., Van Dieijen-Visser, M.P., Wodzig, W.K.: Posttranslational modifications of cardiac troponin T: an overview. J. Mol. Cell. Cardiol. **63**, 47–56 (2013)
49. Huke, S., Knollmann, B.C.: Increased myofilament Ca2+-sensitivity and arrhythmia susceptibility. J. Mol. Cell. Cardiol. **48**, 824–833 (2010)
50. Hart, G.W., Slawson, C., Ramirez-Correa, G., Lagerlof, O.: Cross talk between O-GlcNAcylation and phosphorylation: roles in signaling, transcription, and chronic disease. Annu. Rev. Biochem. **80**, 825–858 (2011)
51. Dubois-Deruy, E., Belliard, A., Mulder, P., Bouvet, M., Smet-Nocca, C., Janel, S., Lafont, F., Beseme, O., Amouyel, P., Richard, V., Pinet, F.: Interplay between troponin T phosphorylation and O-N-acetylglucosaminylation in ischaemic heart failure. Cardiovasc. Res. **107**, 56–65 (2015)
52. Forbes, J.M., Soldatos, G., Thomas, M.C.: Below the radar: advanced glycation end products that detour "around the side": is HbA1c not an accurate enough predictor of long term progression and glycaemic control in diabetes? Clin. Biochem. Rev. **26**, 123–134 (2005)
53. Gugliucci, A., Bendayan, M.: Renal fate of circulating advanced glycated end products (AGE): evidence for reabsorption and catabolism of AGE peptides by renal proximal tubular cells. Diabetologia. **39**, 149–160 (1996)
54. Janssens, J.V., Ma, B., Brimble, M.A., Van Eyk, J.E., Delbridge, L.M.D., Mellor, K.M.: Cardiac troponins may be irreversibly modified by glycation: novel potential mechanisms of cardiac performance modulation. Sci. Rep. **8**, 16084 (2018)
55. Ducheix, S., Magré, J., Cariou, B., Prieur, X.: Chronic O-GlcNAcylation and diabetic cardiomyopathy: the bitterness of glucose. Front. Endocrinol. (Lausanne). **9**, 642 (2018)
56. Yoshioka, K.: Skin autofluorescence is associated with high-sensitive cardiac troponin T, a circulating cardiac biomarker, in Japanese patients with diabetes: a cross-sectional study. Diab. Vasc. Dis. Res. **15**, 559–566 (2018)
57. Tanokura, M., Ohtsuki, I.: Primary structure of chymotryptic subfragments from rabbit skeletal troponin T. J. Biochem. **90**, 263–265 (1981)

58. Tanokura, M., Tawada, Y., Onoyama, Y., Nakamura, S., Ohtsuki, I.: Interactions among chymotryptic troponin T subfragments, tropomyosin, troponin I and troponin C. J. Biochem. **95**, 1417–1421 (1984)
59. Zhang, Z., Biesiadecki, B.J., Jin, J.P.: Selective deletion of the NH2-terminal variable region of cardiac troponin T in ischemia-reperfusion by myofibril-associated μ-calpain cleavage. Biochemistry. **45**, 11681–11694 (2006)
60. Neuhof, C., Neuhof, H.: Calpain system and its involvement in myocardial ischemia and reperfusion injury. World J. Cardiol. **6**, 638–652 (2014)
61. The UniProt Consortium. Nucleic Acids Res. **47**(D1), D506–D515 (2019). Human calpain-1 (CAN1_HUMAN). https://www.uniprot.org/uniprot/P073840. Accessed 17 Nov 2019
62. Yarbrough, W.M., Mukherjee, R., Stroud, R.E., Meyer, E.C., Escobar, G.P., Sample, J.A., Hendrick, J.W., Mingoia, J.T., Spinale, F.G.: Caspase inhibition modulates left ventricular remodeling following myocardial infarction through cellular and extracellular mechanisms. J. Cardiovasc. Pharmacol. **55**, 408–416 (2010)
63. The UniProt Consortium. Nucleic Acids Res. **47**(D1), D506–D515 (2019). Human caspase-3 (CASP3_HUMAN). https://www.uniprot.org/uniprot/P42574. Accessed 17 Nov 2019
64. Communal, C., Sumandea, M., De Tombe, P., Narula, J., Solaro, R.J., Hajjar, R.J.: Functional consequences of caspase activation in cardiac myocytes. Proc. Natl. Acad. Sci. U. S. A. **99**, 6252–6256 (2002)
65. Streng, A.S., De Boer, D., Van Doorn, W.P., Kocken, J.M., Bekers, O., Wodzig, W.K.: Cardiac troponin T degradation in serum is catalysed by human thrombin. Biochem. Biophys. Res. Commun. **481**, 165–168 (2016)
66. Katrukha, I.A., Kogan, A.E., Vylegzhanina, A.V., Serebryakova, M.V., Koshkina, E.V., Bereznikova, A.V., Katrukha, A.G.: Thrombin-mediated degradation of human cardiac troponin T. Clin. Chem. **63**, 1094–1100 (2017)
67. The UniProt Consortium. Nucleic Acids Res. **47**(D1), D506–D515 (2019). Human thrombin (THRB_HUMAN). https://www.uniprot.org/uniprot/P00734. Accessed 17 Nov 2019
68. Streng, A.S., De Boer, D., Van Doorn, W.P., Bouwman, F.G., Mariman, E.C., Bekers, O., Van Dieijen-Visser, M.P., Wodzig, W.K.: Identification and characterization of cardiac troponin T fragments in serum of patients suffering from acute myocardial infarction. Clin. Chem. **63**, 563–572 (2017)
69. Vroemen, W.H.M., Mezger, S.T.P., Masotti, S., Clerico, A., Bekers, O., De Boer, D., Mingels, A.: Cardiac troponin T: only small molecules in recreational runners after marathon completion. J. Appl. Lab. Med. **3**, 909–911 (2019)
70. Diris, J.H., Hackeng, C.M., Kooman, J.P., Pinto, Y.M., Hermens, W.T., Van Dieijen-Visser, M.P.: Impaired renal clearance explains elevated troponin T fragments in hemodialysis patients. Circulation. **109**, 23–25 (2004)
71. Cardinaels, E.P., Mingels, A.M., Van Rooij, T., Collinson, P.O., Prinzen, F.W., Van Dieijen-Visser, M.P.: Time-dependent degradation pattern of cardiac troponin T following myocardial infarction. Clin. Chem. **59**, 1083–1090 (2013)
72. Vroemen, W.H.M., De Boer, D., Streng, A.S., Bekers, O., Wodzig, W.K.W.H.: Thrombin activation via serum preparation is not the root cause for cardiac troponin T degradation. Clin. Chem. **63**, 1768–1769 (2017)
73. Mingels, A.M., Cardinaels, E.P., Broers, N.J., Van Sleeuwen, A., Streng, A.S., Van Dieijen-Visser, M.P., Kooman, J.P., Bekers, O.: Cardiac troponin T: smaller molecules in patients with end-stage renal disease than after onset of acute myocardial infarction. Clin. Chem. **63**, 683–690 (2017)
74. Fahie-Wilson, M.N., Carmichael, D.J., Delaney, M.P., Stevens, P.E., Hall, E.M., Lamb, E.J.: Cardiac troponin T circulates in the free, intact form in patients with kidney failure. Clin. Chem. **52**, 414–420 (2006)
75. Streng, A.S., De Boer, D., Bouwman, F.G., Mariman, E.C., Scholten, A., Van Dieijen-Visser, M.P., Wodzig, W.K.: Development of a targeted selected ion monitoring assay for the elucidation of protease induced structural changes in cardiac troponin T. J. Proteome. **136**, 123–132 (2016)
76. Streng, A.S., De Boer, D., Van Doorn, W.P., Bouwman, F.G., Mariman, E.C., Bekers, O., Van Dieijen-Visser, M.P., Wodzig, W.K.: Mass spectrometric identification of cardiac troponin T in urine of patients suffering from acute myocardial infarction. J. Appl. Lab. Med. **2**, 857–867 (2017)
77. Hall, J.E. Glomular filtration, renal blood flow, and their control. In: Guyton and Hall Textbook of Medical Physiology, Elsevier ClinicalKey, Chapter 27, pp. 335–346 (2016)
78. Van der Linden, N., Cornelis, T., Kimenai, D.M., Klinkenberg, L.J.J., Hilderink, J.M., Lück, S., Litjens, E.J.R., Peeters, F.E.C.M., Streng, A.S., Breidthardt, T., Van Loon, L.J.C., Bekers, O., Kooman, J.P., Westermark, P., Mueller, C., Meex, S.J.R.: Origin of cardiac troponin T elevations in chronic kidney disease. Circulation. **136**, 1073–1075 (2017)
79. Ziebig, R., Lun, A., Hocher, B., Priem, F., Altermann, C., Asmus, G., Kern, H., Krause, R., Lorenz, B., Möbes, R., Sinha, P.: Renal elimination of troponin T and troponin I. Clin. Chem. **49**, 1191–1193 (2003)
80. Kumar Pasupulati, A., Chitra, P.S., Reddy, G.B.: Advanced glycation end products mediated cellular and molecular events in the pathology of diabetic nephropathy. Biomol. Concepts. **7**, 293–309 (2016)
81. Labugger, R., Organ, L., Collier, C., Atar, D., Van Eyk, J.E.: Extensive troponin I and T modification detected in serum from patients with acute myocardial infarction. Circulation. **102**, 1221–1226 (2000)

5 Research Progress of Urine Biomarkers in the Diagnosis, Treatment, and Prognosis of Bladder Cancer

Feng Jin, Muhammad Shahid, and Jayoung Kim

Abstract

Bladder cancer (BC) is one of the most common tumor with high incidence. Relative to other cancers, BC has a high rate of recurrence, which results in increased mortality. As a result, early diagnosis and life-long monitoring are clinically significant for improving the long-term survival rate of BC patients. At present, the main methods of BC detection are cystoscopy and biopsy; however, these procedures can be invasive and expensive. This can lead to patient refusal and reluctance for monitoring. There are several BC biomarkers that have been approved by the FDA, but their sensitivity, specificity, and diagnostic accuracy are not ideal. More research is needed to identify suitable biomarkers that can be used for early detection, evaluation, and observation. There has been heavy research in the proteomics and genomics of BC and many potential biomarkers have been found. Although the advent of metabonomics came late, with the recent development of advanced analytical technology and bioinformatics, metabonomics has become a widely used diagnostic tool in clinical and biomedical research. It should be emphasized that despite progress in new biomarkers for BC diagnosis, there remains challenges and limitations in metabonomics research that affects its translation into clinical practice. In this chapter, the latest literature on BC biomarkers was reviewed.

Keywords

Urine · Biomarkers · Metabolomics · Bladder Cancer

F. Jin · M. Shahid
Departments of Surgery, Cedars-Sinai Medical Center, Los Angeles, CA, USA
e-mail: Peng.jin@cshs.org; Muhammad.Shahid@cshs.org

J. Kim (✉)
Departments of Surgery, Cedars-Sinai Medical Center, Los Angeles, CA, USA

Samuel Oschin Comprehensive Cancer Institute, Cedars-Sinai Medical Center, Los Angeles, CA, USA

Departments of Surgery and Biomedical Sciences, Cedars-Sinai Medical Center, Los Angeles, CA, USA

University of California, Los Angeles, CA, USA

Department of Urology, Gachon University College of Medicine, Incheon, Republic of Korea
e-mail: Jayoung.Kim@cshs.org

5.1 Introduction

5.1.1 Bladder Cancer (BC) Incidence, Epidemiology, and Risk Factors

Bladder cancer (BC) is the fourth most common cancer in the U.S. and the second most common cancer of the urinary system, accounting for 7%

H. M. Baptista Carreira dos Santos (ed.), *Translational Urinomics*, Advances in Experimental Medicine and Biology 1306, https://doi.org/10.1007/978-3-030-63908-2_5

of all new cancer cases. It also accounts for 4% of all cancer-related deaths in the U.S., ranking it the fifth deadliest cancer. The male to female ratio of morbidity and mortality was about 3:1 [1]. Risk factors are related to the environment, diet, and lifestyle, especially smoking, exposure to aromatic amines, and genetic factors [2–4]. Other known risk factors include the ingestion of high levels of arsenic or significant usage of pain relievers containing finazepine [4, 5].

5.1.2 Economic Burden of BC

The European Organization for Research and Treatment of Cancer (EORTC) has established recommended plans for low to moderate-risk BC patients. This involves a cystoscopy every three months during the first two years, every four months during the following two years, and once a year thereafter [6]. Because BC treatment is continuous, the lifetime cost of treatment and monitoring increases with time. Studies have shown that the cumulative cost of health insurance for long-term survivors (those over 16 years) is $172,426 [7]. As a result of this need for lifelong monitoring, the cost per patient when treating BC is the highest of all other cancers [8].

5.1.3 Classical Classification of BC

Based on the degree of invasion in the bladder muscle wall, BC is divided into either non-muscle invasive BC (NMIBC) or muscle invasive BC (MIBC)[9]. There may be different genetic variation underlying the difference between the two types of BC [10]. When histologically subtyping BC, there are several types. Transitional cell carcinoma (TCC), also known as urothelial carcinoma, accounts for about 90% of all BC. Squamous cell carcinoma (SCC) and adenocarcinoma account for about 10%[11]. There are various other rare types of BC as well [12].BC can also be divided pathologically into low-grade (LG) and high-grade (HG) tumors. LG tumors are usually well-differentiated, while HG tumors are poorly differentiated [13].

5.1.4 Molecular Phenotyping of BC

Recent genome mRNA expression analysis demonstrated that BC can be classified into molecular subtypes. These different subtypes of BC have distinct progression patterns, biological and clinical properties, and response to chemotherapies. There are currently five published classification methods; these include guidelines from the University of North Carolina (UNC), MD Anderson Cancer Center (MDA), The Cancer Genome Atlas (TCGA), Lund University (Lund), and Broad Institute of Massachusetts Institute of Technology and Harvard University (Broad) (Table 5.1).

The classifications by UNC define two molecular subtypes of high-grade BC, "luminal" and "basal", with molecular features reflecting different stages of urothelial differentiation [14]. Luminal BC expresses terminal urothelial differentiation markers, such as those seen in umbrella cells (UPK1B, UPK2, UPK3A, and KRT20), whereas basal BC expresses high levels of genes that are typical in urothelial basal cells (KRT14, KRT5, and KRT5B). The UNC study created a gene signature, BASE47, that accurately discriminates intrinsic bladder subtypes. Identified basal tumors had significantly decreased disease-

Table 5.1 Different classifications of BC based on molecular phenotyping. This table does not contain classifications based on Gottfrid's research

UNC	MDA	Lund	TCGA	Broad
Basal	Basal	UroA	Cluster I	Basal
Luminal	Luminal	UroB	Cluster II	Luminal
	P53-like	GU	Cluster III	Luminal immune
		SCCL	Cluster IV	Immune undifferentiated
		Infiltrated		

specific and overall survival. In addition, among the clinicopathological features available in the MSKCC dataset, only subtypes identified by BASE47 were found to be significant in disease-specific survival by univariate analysis. This study also found that females have an increased incidence of basal-like BC, which is associated with a worse prognosis.

The classification system by MDA identified three molecular subtypes of MIBC: "basal", "luminal", and "P53-like" [15]. Basal MIBC was associated with shorter disease-specific and overall survival, presumably because these patients tend to have more invasive and metastatic disease at presentation. Transcription factor P63 plays a central role in controlling basal gene signatures and preliminary data suggests that EGFR, Stat-3, NFκB, and Hif-1α are also involved. Luminal MIBC displays active ER/TRIM24 pathway gene expression and were enriched for FOXA1, GATA3, ERBB2, and ERBB3. Luminal MIBC contains active PPAR gene expression and activating FGFR3 mutations; thereby, PPARγ- and FGFR-3-targeted agents may be active in this subtype. Because luminal MIBC responds well to neoadjuvant chemotherapy (NAC), targeted therapies should be combined with conventional chemotherapy for maximum efficacy. The P53-like MIBC responded very poorly to NAC and were consistently resistant to frontline neoadjuvant cisplatin-based combination chemotherapy. Additionally, comparative analysis of matches gene expression profiles before and after chemotherapy revealed that all resistant tumors expressed wild-type P53 gene expression signatures. These results indicate that "P53-ness" may play a central role in BC chemoresistance.

The classification by TCGA identified four clusters (clusters I–IV) by analyzing RNA-seq data from 129 tumors [16]. Cluster I (papillary-like) is enriched in tumors with papillary morphology, FGFR3 mutations, FGFR3 copy number gain, and elevated FGFR3 expression. Cluster I samples also had significantly lower expression of miR-99a, miR-100, miR-145 and miR-125b. Tumors with FGFR3 alterations and those that share similar cluster I expression profiles may respond well to inhibitors of FGFR and its downstream targets. Clusters I and II express high levels of GATA3 and FOXA1. Markers of urothelial differentiation, such as uroplakins, epithelial marker E-cadherin, and members of miR-200 miRNAs are also highly expressed in clusters I and II. Clusters I and II express high HER2 levels and an elevated estrogen receptor beta signaling signature, which suggests potential targets for hormone therapies, such as tamoxifen or raloxifene. Cluster III (basal/squamous-like) express characteristic epithelial lineage genes, including KRT14, KRT5, KRT6A, and EGFR. Many of the samples in cluster III express cytokeratins (KRT14 and KRT5). Integrated expression profiling analysis of cluster III revealed a urothelial carcinoma subtype with cancer stem-cell expression features, perhaps providing another avenue for therapeutic targeting.

The Lund classification system defines five major urothelial carcinoma subtypes: urobasal A, genomically unstable, urobasal B, squamous cell carcinoma-like (SCC-like), and infiltrated tumor class [17]. This was established using gene expression profiles from 308 tumor cases. These different molecular subtypes show significantly different prognosis. The best prognosis is the urobasal A, and the worst prognosis are urobasal B and SCC-like. The prognosis of genomically unstable and infiltrated class are between them. Urobasal A tumors were characterized by elevated expression of FGFR3, CCND1, TP63, as well as expression of KRT5 in cells at the tumor–stroma interface. The majority of urobasal A tumors were non-muscle invasive and of low pathologic grade. The genomically unstable subtype was characterized by expression of ERBB2 and CCNE, low expression of cytokeratin, and frequent mutations of TP53. Genomically unstable cases represented a high-risk group, as close to 40% were MIBC. This subtype also showed low PTEN expression. The SCC-like subtype was characterized by high expression of basal keratins, which are normally not expressed in the urothelium; these include KRT4, KRT6A, KRT6B, KRT6C, KRT14, and KRT16. SCC-like tumors also had markedly bad prognoses. Furthermore, this group showed a comparatively different proportion of female/male patients, reminiscent of the 1:1 proportion seen in patients diagnosed with bladder SCC, suggesting that females are more likely to develop urothelial carcinomas with a keratinized/squamous phenotype, which is associated with an adverse prognosis. Urobasal B tumors showed several similarities to urobasal A tumors, such as a high FGFR3 mutation frequency, elevated FGFR3, CCND1, and TP63 levels, and expression of the FGFR3 gene signature. However, this group also showed frequent TP53 mutations and expression of several keratins specific for the SCC-like subtype. Additionally, 50% of the cases were MIBC; including 5of 9 FGFR3 mutated cases. The infiltrated subtype demonstrated a pronounced immunologic and extracellular membrane (ECM) signal, indicating the presence of immunologic and myofibroblast cells. This subtype most likely represents a heterogeneous class of tumors; immunohistochemistry (IHC) revealed the presence of tumors with genomically unstable, urobasal B, and SCC-like protein expression patterns in this group.

The Broad classification identified four different subtypes: luminal, immune undifferentiated, luminal immune, and basal [18]. Approximately 41% of invasive BC was in the luminal subtype, with high expression of KRT20 and UPKs 2/1A/1B/3A as well as moderate to high expression of multiple pertinent transcription factors (KLF5, PPARG, and GRHL5). The luminal subtype was enriched for in male patients, papillary histology, and stage II tumors. A third (29%) of invasive BC was in the basal subtype, with high expression of KRT14, KRT5, KRT6A/B, and KRT16, and low expression of uroplakins, which is consistent with basal or undifferentiated cytokeratin expression patterns. Consistent with prior studies, the basal subtype expressed TP63, TP73, MYC, EGFR, TGM1, and SCEL, which is indicative of some degree of squamous differentiation. The basal subtype was enriched in female patients and tumors with nonpapillary histology. The basal subtype also expressed many immune genes at intermediate and somewhat variable levels. These genes include CTLA4 and CD274, which encodes for PD-L1, suggesting that there may be immune cell infiltration of tumors. A smaller percentage of cancers (11%) were grouped into a novel subtype called immune undifferentiated. These cancers showed very low expression of luminal markers, variable expression of basal cytokeratins, and relatively high expression of immune genes, including CTLA4 and CD274, which further suggests significant immune cell infiltration and possible immune evasion. Lastly, the luminal immune subtype group constitutes about 18% of all cases and is characterized by the expression of luminal genes (cytokeratins and uroplakins) and intermediate expression of immune genes. This group was notably enriched for stage N+ tumors. The luminal subtype was enriched for in cancers with FGFR3 mutations and amplification events involving PVRL4 and YWHAZ. The basal subtype was enriched for NFE2L2 mutations. Both the luminal immune and immune undifferentiated subtypes had high expression levels of ZEB1, ZEB2, and TWIST1, which is characteristic of epithelial-mesenchymal transition (EMT).

Gottfrid et al. proposed five major tumor-cell phenotypes in advanced BC: urothelial-like, genomically unstable (GU), basal/SCC-like, mesenchymal-like, and small-cell/neuroendocrine-like [19]. Urothelial-like tumors express FGFR3 and CCND1 and frequently demonstrate a loss of 9p21 (CDKN2A). GU tumors express FOXM1, but not KRT5, and frequently show loss of RB1. Basal/SCC-like tumors were found to express KRT5 and KRT14, but not FOXA1 and GATA3. The mesenchymal-like BC is a new subtype that shows a tumor-cell phenotype that starkly contrasts with previously defined subtypes and is biologically different from the basal/SCC-like cases that they are clustered with. The tumor cells are mesenchymal-like and express typical mesenchymal genes, such as ZEB2 and VIM. The tumor cells were themselves mesenchymal-like and expressed the typical mesenchymal genes ZEB2 and VIM. The consensus cluster Sc/NE-like turned out to harbor two very distinct tumor-cell phenotypes. One-half of these tumors expressed markers that are typical for neuroendocrine differentiation. This part of the Sc/NE consensus cluster also showed an absence of PPARG, FOXA1, and GATA3 expression, as well as of uroplakin and KRT20 expression.

Kardos et al. reported the discovery of a claudin-low molecular subtype of high-grade BC that shares characteristics with the homonymous subtype of breast cancer [20]. Although there has been much work done on the molecular phenotyping of BC, the different emphases of different classification methods have made it difficult to consolidate a widely accepted classification method. As a result, the molecular phenotyping of BC remains to be further studied. The claudin-low subtype can be considered a subpopulation of the basal-like subtype (UNC classification system). Claudin-low bladder tumors are rich in a variety of genetic characteristics, including increased mutation rates of RB1, EP300, and NCOR1, increased the frequency of EGFR amplification, decreased mutation rates of FGFR3, ELF3, and KDM6A, and decreased the frequency of PPARG amplification. These characteristics define a molecular subtype of BC with distinct molecular features and an immunological profile that is theoretically primed for an immunotherapeutic response.

Figure 5.1 summarizes the classification of BC.

5.2 Biomarker Discovery in BC

More than 75% of patients are diagnosed and treated for NMIBC. At the time of initial evaluation, its recurrence rate can be as high as 70% [21]. Currently, the standard and most important examination method for BC is cystoscopy, However, this procedure is invasive, uncomfortable, and expensive [22]. Furthermore, cystoscopy may miss certain lesions, particularly smaller areas of carcinoma in situ [23]. Molecular biosignatures indicative of altered cellular landscapes and functions have been casually linked to pathological conditions, suggesting the promise of BC-specific biomarkers. However, a noninvasive biomarker that is as sensitive and specific as standard cystoscopy has yet to be discovered. As we progress through the twenty-first century, we now have access to a number of ways to analyze diagnostic markers in-depth. The evolution of omics platforms and bioinformatics to allow for analysis of the genome, epigenome, transcriptome, proteome, lipidome, metabolome et al. enables the development of more sensitive biomarkers. These discoveries will broaden under-

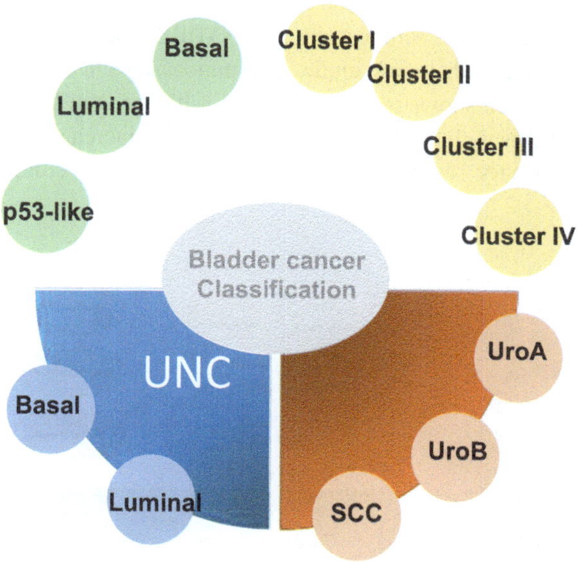

Fig. 5.1 Schematic illustration of molecular subtypes of bladder cancer. Based on Whole-genome mRNA expression profiling, several molecular subtypes of muscle-invasive bladder cancer (MIBC) have been identified. Molecular subtypes of MIBC might have important implications for patient prognosis and response to conventional chemotherapy and targeted agents. Four groups have shown great similarities among tumor subtype. Lund, University of Lund; MDA, MD Anderson Cancer Center; TCGA, The Cancer Genome Atlas; UNC, University of North Carolina

standing of the complex biology and pathophysiology of bladder diseases, which can then be clinically translated. Biomarkers of interest can be detected in different types of samples, including serum, tissue, and urine. Urinary biomarkers are particularly attractive due to cost, time, and minimal effort. As a result, studies on urinary BC biomarkers continue to expand.

Figure 5.2 shows the overview of the multi-OMICS strategies for urine-based biomarker discovery and translational application.

5.2.1 Proteomics-Based BC Biomarkers

In patients with hematuria, aurora A kinase (AURKA) can discriminate low-grade BC patients vs. normal patients [24]. After adjusting for patients, clinical features, and treatment with Bacillus Calmette-Guerin, the activated leukocyte cell adhesion molecule (ALCAM) is positively correlated with tumor stage and overall survival (OS) [25]. Nicotinamide N-methyltransferase has been shown to be elevated in BC patients and is correlated with histological grade [26]. Apurinic/apyrimidinic endonuclease 1/redox factor-1 (APE/Ref-1) levels are higher in BC, with respect to non-BC, and is correlated with tumor grade and stage; moreover, it has been shown to be significantly increased in patients with historical BC recurrence [27]. The urinary cytokeratin-20 (CK20) RT-PCR assay shows that the sensitivity of urothelial BC detection was 78–87%, and the specificity was 56–80%., with improved diagnostic accuracy in tumor progression [28]. However, its performance is relatively poor in low-grade tumors. Higher urinary levels of CK8 and CK18 have been detected via UBC Rapid Test in high vs. low-grade BC [29].

There are multiple markers that can potentially be used for BC detection; increased urinary levels of apolipoproteins, A1, A2, B, C2, C3, and E (APOA1, APOA2, APOB, APOC2, APOC3, APOE) were found in BC compared to healthy controls [30, 31]. A signature of 4 urinary fragments of uromodulin, collagen α-1 (I), collagen α-1 (III), and membrane-associated progesterone receptor component 1 may be able to discriminate MIBC from NMIBC [32]. Other panels employ IL-8, MMP-9/10, ANG, APOE, SDC-1, α1AT, PAI-1, VEGFA, and CA9 to indicate BC from urine samples. The advantage of these multi-urinary protein biomarkers is evident in high and low-grade and high and low-stage diseases [33]. Combined with urine markers, including midkine (MDK), MDK, synuclein G, CEACAM1, ZAG2 [34], clusterin (CLU) and angiogenin (ANG), the sensitivity and specificity of NMIBC diagnosis can be improved through immunoassay and urine cytology [35]. CK20 and insulin-like growth factor II (IGF-II) levels were found to be increased in the urine sediments of NMIBC patients compared to controls [36].

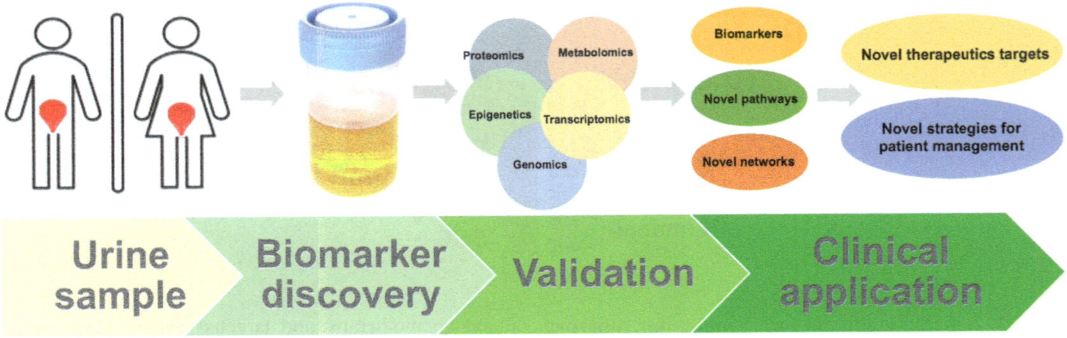

Fig. 5.2 Overview of the multi-OMICS strategies for urinary bladder cancer biomarker discovery and their clinical implication. A typical integrated multi-omic technologies workflow showing to probe the complexity of bladder cancer biology. Integration of several of omics data sources use systems biology approach build biomarker discovery

Table 5.2 Summary of proteomics-based BC biomarkers

Samples	Proteins	References
Urine	AURKA	[24]
	ALCAM	[25]
	Nicotinamide N-methyltransferase	[26]
	APE/Ref-1	[27]
	CK20	[28]
	CK8, CK18	[29]
	APOA1, APOA2, APOB, APOC2, APOC3, APOE	[30, 31]
	Uromodulin, collagen α-1 (I), collagen α-1 (III), and membrane-associated progesterone receptor component 1	[32]
	IL-8, MMP-9/10, ANG, APOE, SDC-1, α1AT, PAI-1, VEGFA, and CA9	[33]
	Midkine (MDK), synuclein G or MDK, ZAG2, CEACAM1 adn angiogenin, clusterin	[34, 35]
	CK20, IGFII	[36]
	HAI-1, Epcam	[37]
	Survivin	[38]
	Snail	[39]
	CD44	[40]

Increased levels of urinary HAI-1 and epithelial cell adhesion molecule (EpCAM) are prognostic biomarkers in high-risk NMIBC patients [37]. Urine survivin have been proved by chemiluminescence enzyme immunoassay that it is a potential biomarker for BC, which has been shown to be related to tumor stage, lymph node metastasis, and distant metastasis.[38]. Snail overexpression represents an independent prognostic factor for tumor recurrence in NMIBC [39]. CD44 in urine was found to be elevated in high-grade MIBC by glycan-affinity glycoproteomics nanoplatforms. [40].

Proteomics-based BC biomarkers were summarized in Table 5.2.

5.2.2 Metabolomics-Based BC Biomarkers

Urinary metabolomics signature may be useful in detecting early stage BC. Jin X et al. analyzed urinary metabolites by high-performance liquid chromatography-quadrupole time-of-flight mass spectrometry (HPLC-QTOFMS), and found 12 metabolites that help to identify BC. [41]. Zhou Y et al. developed a urinary pseudotargeted method based on gas chromatography-mass spectrometry (GC-MS) which has been validated by a BC metabolomics study [42]. Using binary logistic regression analysis, a four-biomarker panel was defined for the diagnosis of BC. The results revealed that the urinary four-biomarker panel can be used to diagnose NMIBC or low-grade BC. Among the four metabolites, cholesterol levels were significantly increased in the BC group, while 5-hydroxyvaleric acid, 3-phosphoglyceric acid, and glycolic acid levels were markedly decreased in the BC group.

X. Cheng et al. carried out a study based on metabolomics with liquid chromatography-high resolution mass spectrometry (LC-HRMS) to discover novel biomarkers for detecting early-stage BC. [43]. A total of 284 subjects were enrolled in the study including 117 healthy adults and 167 BC patients. Metabolite panels are known to have more predictive power than a single metabolite [44]. A metabolite panel consisting of dopamine 4-sulfate, MG00/1846Z,9Z,12Z,15Z/00, aspartyl-histidine, and tyrosyl-methionine was found to have the best predictive accuracy in diagnosing NMIBC.

A study by Yumba Mpanga A et al. developed and validated an analytical method for the simultaneous quantitative determination of metabolites using reversed phase high-performance liquid chromatography coupled with triple quadrupole mass spectrometry (RP-HPLC-QQQ/MS) [45]. The optimized and validated method was applied to urine samples from 40 BC patients and 40 healthy matched controls. Statistical analysis was done using the Student's t-test or U-Mann Whitney test. This identified 10 compounds that participate in different metabolic pathways, such as gut flora metabolism, RNA degradation, purine metabolism, etc., as being significantly different in urine between BC and control groups ($p < 0.05$). These 10 compounds include acetyl-lysine, N-acetylneuraminic acid, pseudouridine, uridine, xanthine, 7-methylguanine, gluconic

Table 5.3 Summary of metabolomics-based BC biomarkers

Metabolites	Alteration	References
Succinate	↑	[41]
Pyruvated	↑	
Oxoglutarated	↑	
Carnitine	↑	
Phosphoenolpyruvate	↑	
Trimethyllysine	↑	
Melatonin	↓	
Isovalerylcarnitine	↑	
Glutarylcarnitine	↓	
Octenoylcarnitine	↑	
Decanoylcarnitine	↑	
Acetyl-CoA	↑	
Cholesterol	↑	[42]
5-hydroxyvaleric acid	↓	
3-phosphoglyceric acid	↓	
Glycolic acid	↓	
Dopamine 4-sulfate	↑	[43]
MG00/1846Z,9Z,12Z,15Z/00	↓	
Aspartyl-histidine		
Tyrosyl-methionine		
Acetyllysine	↑	[45]
N-acetylneuraminic acid	↑	
Pseudouridine	↑	
Uridine	↑	
Xanthine	↑	
7-methylguanine	↑	
Gluconic acid	↑	
Glucuronic acid	↑	
1,7 dimethylxanthine	↓	
Hippuric acid	↓	
Acid trehalose	↑	[46]
Nicotinuric acid	↑	
AspAspGlyTrp peptide	↑	
Inosinic acid	↓	
Ureidosuccinic acid	↓	
GlyCysAlaLys peptide	↓	

acid, glucuronic acid, 1,7 dimethylxanthine, and hippuric acid. Moreover, acid trehalose, nicotinic acid, and AspAspGlyTrp peptide were upregulated; inosinic acid, ureidosuccinic acid, and GlyCysAlaLys peptide were downregulated in BC, but not in healthy controls [46].

Metabolomics-based BC biomarkers were summarized in Table 5.3.

5.2.3 Genomics-Based BC Biomarkers

5.2.3.1 DNA Methylation

Using urine sediments from BC patients, Sun and her group demonstrated that SOX-1, IRAK3, and Li-MET gene methylation status have higher recurrence predictivity than urine cytology and cystoscopy (80 vs. 35 vs. 15%, respectively) [47]. Methylated genes, such as those for APC and cyclin D2, were found to be significantly prevalent in the urine from malignant vs. benign cases [48]. Hypermethylation of the GSTP1 and RARβ2 and APC genes have been identified in the urine of BC patients [49]. Evaluation of Twist Family BHLH Transcription Factor 1 (TWIST1) and NID2 genes methylation status in urine has been shown to differentiate primary BC patients from controls with 90% sensitivity and 93% specificity [50]. Additionally, evaluation of the methylation status of NID2, TWIST1, CFTR, SALL3, and TWIST1 genes in urinary cells in combination with urine cytology has been found to increase sensitivity and have high negative predictive value in BC patients [51, 52]. Urinary methylation levels of POU4F2 and PCDH17 is able to distinguish BC from normal controls with 90% sensitivity and 94% specificity [53]. Promoter hypermethylation of HS3ST2, SEPTIN9, and SLIT2 combined with FGFR3 mutation showed 97.6% sensitivity and 84.8% specificity in the diagnosis, surveillance, and risk stratification of low- and high-risk NMIBC patients [54]. Lastly, the methylation status of p14ARF, p16INK4A, RASSf1A, DAPK, and APC has been found to be correlated with BC grade and stage [55].

5.2.3.2 miRNAs

Urinary levels of miR-146a-5p are significantly increased in high-grade BC [56]. MiR-126 urinary levels were found to be elevated in BC compared to healthy controls [57]. Low miR-200c expression has been shown to be correlated with tumor progression in NMIBC [58]. Chen et al. detected 74 miRNAs, of which 33 were upregulated and 41 were downregulated in BC compared to healthy patients; the most notable of these include let-7miR, mir-1268, miR-196a, miR-1, miR-100, miR-101, and miR-143 [59]. By screening patients with negative cystoscopy, Eissa et al. identified miR-96 and miR-210 as being associated with BC [60]. MiR-125b, miR-30b, miR-204, miR-99a, and miR-532-3p were downregulated in the urine supernatant of BC patients [61]. MiR-9, miR-182, and miR-200b have been shown to be correlated with MIBC aggressiveness, recurrence-free, and overall survival (OS) [62]. MiR-145 distinguishes NMIBC from non-BC [63]. MiR-144-5p inhibits BC proliferation, affecting CCNE1, CCNE2, CDC25A, and PKMYT1 target genes [64]. Cell-free urinary miR-99a and miRNA-125b were found to be downregulated in the urine supernatants of BC patients (sensitivity 86.7%; specificity 81.1%) [65]. Urinary levels of miR-618 and miR-1255b-5p were increased in MIBC patients compared to healthy controls [66]. Whole genome analysis determined increased miR-31-5p, miR-191-5p and miR-93-5p levels in the urine of BC patients compared to controls [67].

Genomics-based BC biomarkers were shown in Table 5.4.

5.3 Metabolomics and Metabolic Phenotypes of BC

In biological research, the omics approach includes genomics, proteomics, and metabolomics. It probes physiological and malignant processes at the cellular and molecular levels; thereby, characterizing the global molecular quantity, structure, function, and dynamic changes within an organism. Although genomics and proteomics have helped subtype many cancers based on gene mutation or receptor status, considerable heterogeneity is observed in tumor behavior and patient outcome, even within a genomic subtype. This is due to the unique cellular processes and metabolic profiles that can only be elucidated through metabolomics [68]. Metabolomic analysis is less complex compared to genomics, transcriptomics, and proteomics due to fewer endpoints. Metabolomics measures the entire set of small molecule products of metabolic processes in a biological system. By focusing on the downstream products of genomic and proteomic processes, metabolomics summarizes the effects of other omics methods and most closely represents a system's phenotype [69].

Metabolomic studies are either untargeted, aiming to comprehensively include all measurable analytes without a prior hypothesis, or targeted, measuring only select predefined groups of metabolites [70]. Although untargeted studies deal with large complex data sets and carry the risk of false positives due to multiple testing of variables, the advantage is that they are free from assumptions. Targeted studies, on the other hand, are hypothesis-driven and offer measurements of high precision and accuracy. In metabolomic biomarker research, targeted studies are often used to validate findings from prior untargeted studies [71].

The field of blood-based genomic and proteomic cancer biomarkers are more developed than that of urine-based metabolomics because blood is considered to be an active participant in biological processes unlike urine, which is a contrast to waste product. With the advancement of urine analysis technology, urinalysis techniques have improved considerably. There are a number of methods that now enable in-depth analysis of diagnostic markers. In particular, NMR and MS-based identification of urinary metabolites are powerful techniques that can potentially diagnose a number of conditions. At present, urine metabolomic biomarker studies are being primarily conducted by either NMR or MS. Both of these tools have their strengths and limitations. The major advantage of MS is its accuracy and specificity in regard to metabolite detection. MS is more accurate compared to NMR spectrometry; however, the analytes need to be separated for detection and assimilation. In contrast, NMR-based spectrometry is more expensive and has lower sensitivity, generally limited to less than 100 analytes in biological fluids. Furthermore, NMR does not require the segregation of analytes for detection. The major advantage of NMR is that samples are not destroyed and can actually be reused [72–74].

Table 5.4 Summary of genomics-based BC biomarkers

	Biomarkers	Alteration	References
DNA methylation	SOX-1, IRAK3, and Li-MET		[47]
	APC and cyclin D2		[48]
	GSTP1 and RARβ2 and APC		[49]
	TWIST1 and NID2		[50]
	NID2 and TWIST1 or CFTR, SALL3 and TWIST1		[51, 52]
	POU4F2 and PCDH17		[53]
	HS3ST2, SEPTIN9 and SLIT2		[54]
	p14ARF, p16INK4A, RASSF1A, DAPK, and APC tumor suppressor		[55]
miRNAs	miR-146a-5p	↑	[56]
	MiR-126	↑	[57]
	miR-200c	↓	[58]
	Let-7miR, mir-1268, miR-196a, miR-1, miR-100, miR-101, and miR-143		[59]
	miR-96 and miR-210		[60]
	MiR-125b, miR-30b, miR-204, miR-99a, and miR-532-3p	↓	[61]
	MiR-9, miR-182 and miR-200b		[62]
	MiR-145		[63]
	MiR-144-5p		[64]
	miR-99a and miRNA-125b	↓	[65]
	miR-618 and miR-1255b-5p	↑	[66]
	miR-31-5p, miR-191-5p and miR-93-5p	↑	[67]

BC has profound metabolic abnormalities. Several altered metabolic pathways play a role in bladder tumorigenesis. As a result, metabolomics can contribute substantially to understanding the relevant alterations of catabolic and anabolic metabolic processes impaired in cancer through the identification of tumor-specific metabolic biomarkers with potential diagnostic, prognostic, or predictive value [75]. Metabolomic studies have already identified various metabolites of diverse pathways (glucose, lipid, amino acid, nucleotide metabolites) as probable BC biomarkers [76].

However, caution must be applied; clinical metabolic phenotypes (metabotypes) may be altered due to age, gender, diet, race, lifestyle, surgical intervention, and underlying pathophysiological conditions [77]. In the context of BC metabolomics, baseline characteristics, such as tumor stage and grade, hematuria (gross or micro), surgical interventions, and smoking habit should additionally be taken into consideration [78].

5.4 Metabolomic Platforms

Contrary to the genome or proteome, the human metabolome composition is still not fully defined. There are few research approaches, all of which have emerged in metabolome analysis; these include metabolic profiling, metabolic fingerprinting and metabolic footprinting [79].

Metabolic profiling is an example of a targeted approach, focusing on identifying and quantifying predetermined groups of metabolites with similar physicochemical properties (e.g., carbohydrates, amino acids, organic acids, nucleosides) or under the same biochemical pathway (e.g., glycolysis, gluconeogenesis, β-oxidation or citric acid cycle) [80]. Metabolic profiling is considered to be an extension of metabolite targeted analysis, which relies on analyzing a single compound or small subset of metabolites to determine the influence of the specific stimuli on metabolism. Metabolic fingerprinting is an untargeted approach that is not driven by any preliminary assumption and aims to define changes in the whole metabolome, which occurs at a specific state in the cell, tissue or organism. Therefore, the main purpose of metabolic fingerprinting is to identify and qualify as many possible metabolites in samples. Metabolic fingerprinting is frequently used in a comparative analysis of two subject groups (i.e., healthy vs disease, one disease vs another disease), which makes it a promising tool in studies focused on disease diagnosis and prognosis [81]. Metabolic footprinting is often applied in microbiological or biotechnological studies. Compared to the other methods, this approach does not concern intracellular metabolites but focuses on compounds that are secreted or failed to be used by cells in specific media. Due to the close relationship between intracellular and extracellular metabolism, metabolic footprinting can provide an integrative interpretation of the metabolic network in a specific living system [82].

Due to both the physicochemical diversity of the metabolome and complexity of the biological systems, no single analytical platform is able to determine all metabolites present in complex biofluids. Therefore, numerous analytical platforms are commonly used in both targeted and untargeted metabolomic studies [83]. NMR or MS coupled with different separation techniques currently dominates in metabolomics. There are at least four major analytical platforms with proven utility for metabolomic applications: NMR, GC-MS, LC-MS, and LCECA [84]. Each of these platforms has specific advantages and disadvantages (Table 5.5).

Modern NMR makes it possible to perform rigorous structural analysis of many metabolites in crude extracts, cell suspensions, intact tissues, or whole organisms. Structural determination of known metabolites using various one-dimensional (1D) or 2D NMR methods is straight forward; in fact, de novo structural analysis of unanticipated or even unknown metabolites is also feasible. NMR has high throughput capability and is particularly capable of determining the structure of metabolites, including the location of isotope labeled atoms in different isotopes produced during stable isotope tracing studies [85–88]. As a result, metabolic pathways can now be systematically mapped by NMR with unprecedented

Table 5.5 Summary of the advantages and limitations of different metabolomics platforms

	Strengths	Drawbacks
NMR	Rapid	Lack of sensitivity
	Reproducible	Multiplicity of the resonance
	Nondestructive	Difficulty of quantification-chemical noise and signal overlapping
	High-throughput	Lack of an analyte separation component
	Minimal sample	High instrument cost (over one million dollars)
	Manipulation	
	Possible tissue analysis	
MS	High sensitivity	Low quantitation
	Wide detection range	Low reproducibility
	Easy metabolite	Destructive
	Identification-databases availability	High sample volume requirements
	Possibility to couple with separation techniques	
GC-MS	Reasonable quantitative precision	Can't study nonvolatile molecules
	High throughput	Low mass accuracy (often unit resolution)
	Low instrumentation costs ($100–$300,000)	Undesirable metabolite losses
	High sensitivity	
	Volatile and thermally stable analytes	
LC-MS	High flexibility	High instruments cost($100,000-over one million dollars)
	Tailor separations to the compounds	Difficulty in obtaining consistent quantitative precision
	Enable low, medium, or high mass accuracy	
	Can trade off sensitivity for throughput	

(continued)

Table 5.5 (continued)

	Strengths	Drawbacks
	Can determine the exact molecular composition	
	Various biofluids analytes	
CE–MS	Highly polar ionogenic metabolites analytes	Notable migration time shift during analyses
	Minimal sample preparation	
	High resolution power	
LCECA	Extremely sensitive	Lack of structural information
	Strong run-to-run precision	Low throughput
	High specificity (tryptophan and tyrosine pathways)	Low cost (under $100,000)

speed. In summary, NMR offers essentially universal detection, excellent quantitative precision, and the potential for high-throughput (>100 samples/day is possible). NMR is an unbiased, robust, reproducible, non-destructive and selective analytical platform. In NMR analysis almost no sample pretreatment is required. However, the main disadvantages of this technique include low sensitivity and lack of analyte separation. Another disadvantage is its high initial cost; NMR instruments can cost well over a million dollars.

MS represents a universal, sensitive tool that can be used to characterize, identify, and quantify a large number of compounds in biological samples where metabolite concentrations may constitute a broad range [89]. Liquid chromatography coupled with mass spectrometry (LC–MS), gas chromatography coupled with mass spectrometry (GC–MS) or capillary electrophoresis coupled with mass spectrometry (CE–MS) has a significantly wider application in metabolome analysis [83].

GC, which employs high-resolution capillary columns and is combined with MS detection, is a powerful platform for determining the metabolome. GC–MS often employs either an electron impact (EI) or chemical ionization (CI) mode, which provides putative identification based on the highly reproducible mass spectra of metabolites and availability of universal structural and mass spectral libraries [90]. GC-MS can provide structural information (more informative if the compounds are present in existing libraries), reasonable quantitative precision, and high-throughput (>100 samples/day is possible). Sensitivity is at least 2 orders of magnitude higher than NMR. One limitation of GC-MS is its inability to study molecules that cannot be readily volatilized. Another limitation is its relatively low mass accuracy (unit resolution). GC-MS is a technique of choice for volatile and thermally stable analytes. Therefore, complex and time-consuming sample derivation is necessary; however, this can lead to undesirable metabolite loss. The recent development of multidimensional GC (GC x GC) has improved resolution, robustness, and sensitivity compared to conventional GC-MS.

LC–MS is the most suitable technique for analyzing non-volatile, thermally unstable, high or low-molecular-weight compounds with a wide polarity range. Most compounds can be analyzed by LC-MS. LC–MS is commonly used in the metabolomic analysis of various biofluids (urine, blood or tissue extracts) [91, 92]. One limitation of LC-MS is its relative difficulty in obtaining consistent quantitative precision. The development of the LC–NMR-MS systems combines the high-throughput capability of NMR with the high sensitivity and resolution of LC–MS [93, 94]. To improve the sensitivity of conventional LC–MS technique, nanoLC–MS was implemented in metabolomics studies [95, 96].

Compared to LC–MS or GC–MS, CE–MS is rarely applied in metabolomic studies. However, recent significant improvements have opened CE-MS application in metabolomics. This technique is particularly useful in analyzing highly polar ionogenic metabolites in biological fluids [97]. CE-MS is a suitable method for urinary metabolomic analysis, which can be performed with relatively minimal sample preparation. However, extensive research is also being conducted in applying CE-MS to serum metabolomics [98]. CE-MS is a technique dedicated to water-soluble and charged molecules, which makes it a highly complementary platform to other separation methods, like LC-MS or GC-MS. The main advantages of CE-MS include high resolution power and small sample or reagent requirements. Its main limitation is the unstable electroosmotic flow phenomenon, which can result in notable migration time shifts during analyses [99].

LCECA is ideal for studies on the tryptophan and tyrosine pathways that lead to monoamine neurotransmitters because many metabolites within these two pathways can be measured quantitatively with LCECA. The robust nature of this platform, its reproducibility, and sensitivity have been well described in a series of peer-reviewed publications [100–104]. Preliminary experiments described later in this review demonstrate the power and promise of electrochemistry-based platforms for metabolomics analysis in defining signatures for central nervous system (CNS) disorders and treatments. The LCECA system is extremely sensitive, perhaps 2–3 orders of magnitude higher than that of GC-MS, and displays strong run-to-run precision over long periods of time. The disadvantages include the lack of structural information and low throughput (12 samples/day is the most commonly used metabolomic configuration). The system can detect molecules, such as tyrosine and tryptophan metabolites, as well as antioxidants and oxidative damage products, but it is "blind" to other molecules, such as glucose, ketoglutarate, and most fatty acids.

Table 5.5 shows the advantages and limitations of different metabolomics platforms.

5.5 Metabolomics in BC Diagnosis and Prognosis and Predicting Response to Therapies

BC has profound metabolic anomalies that play central roles in tumor progression [105]. Metabolic pathways, such as the tricarboxylic acid (TCA) cycle, lipid synthesis, amino acid synthesis, nucleotide synthesis, and glycolysis pathway, are known to be increased in BC tissue compared to adjacent benign tissue [106].

5.5.1 Tricarboxylic Acid Cycle

A significant decrease in citrate concentration was consistently observed in the urine and serum of BC patients [107]. One possible explanation for this is the active uptake of citrate from the extracellular medium into the tumor cell [108]. Citrate is important for lipid biosynthesis, which is crucial for tumor proliferation [109]. Therefore, the decrease in citrate levels in the urine or serum may illustrate the increased utilization of citrate in lipogenesis for the rapid proliferation of tumor cells [2].

5.5.2 Lipid Metabolism

Up or downregulation of carnitine species, including carnitine, carnitine C8:1, carnitine C9:0, carnitine C9:1, carnitine C10:1, carnitine C10:3, isobutyryl carnitine, acetylcarnitine, 2,6-dimethylheptanoylcarnitine, isovalerylcarnitine, glutarylcarnitine, and decanoylcarnitine, has been reported in BC [41, 110, 111]. The carnitine system plays a central role in lipid metabolism; it facilitates the entry of long-chain fatty acids into the mitochondria for utilization in energy-generating processes and removes short-chain and medium-chain fatty acids that accumulate as a byproduct [112]. It has been postulated that the dysregulation of lipid metabolism provides an environment that is beneficial to the development of BC. Additionally, altered fatty acid transportation, fatty acid β-oxidation, or energy metabolism might partially explain why BC patients are prone to lethargy [2].

5.5.3 Amino Acid Metabolism

5.5.3.1 Glutathione Metabolism

Elevated glutathione (GSH) level was reported in BC tissues and cell lines via metabonomic studies [2]. Oxidative stress results in elevated GSH and overexpression of antioxidant enzymes, such as glutathione peroxidase, glutathione reductase, and glutathione-S transferase [113]. While GSH is involved in the detoxification of carcinogens, its elevation in tumors may promote chemotherapy resistance in cancer cells via conjugation with pharmacologically active drugs or metabolites [114].

5.5.3.2 Tryptophan Metabolism

Upregulation of tryptophan metabolism in BC was observed with increased levels of anthranilic acid, N-acetylanthranilic acid, kynurenine, 3-hydroxykynurenine, and malonate [115–117]. The proposed underlying mechanisms include autoxidation and interaction with nitrite or transition metals to form reactive intermediates, binding as ligands to aryl hydrocarbon receptor (AHR) that plays a role in carcinogenesis [118]. Notably, Opitz et al. demonstrated that tryptophan-2,3-dioxygenase (TDO)-derived kynurenine suppresses antitumor immune responses and promotes tumor-cell survival through AHR, which in turn suggests TDO as a potential cancer therapeutic target [119].

5.5.3.3 Hippuric Acid & Taurine Metabolism

Downregulation of hippuric acid was generally observed in BC patients and taurine was found to be elevated in BC patients compared to benign and healthy controls [107]. Taurine inactivates hypochlorous acid, which is a strong oxidant and cytotoxic agent, by forming stable taurine chloramine (Tau-Cl). In turn, Tau-Cl downregulates immunological responses via production of pro-inflammatory cytokines, leading to tumor progression [120].

5.5.3.4 Nucleotide Metabolism

Purine and pyrimidine metabolism has been found to be perturbed in BC, leading to upregulation of guanine, hypoxanthine, cytidine monophosphate, thymine, uracil, uridine, and pseudouridine [111, 115]. Nucleosides, particularly modified nucleosides (e.g., pseudouridine), are elevated and suggested as potential biomarkers in various cancers [121]. Such elevation nucleoside levels have been postulated to be the result of increased DNA synthesis associated with enhanced cell cycle activity in cancer [122]. Modified nucleosides are excreted in urine because they cannot be recycled as nucleosides [123]. Thus, levels of modified nucleosides in urine reflect oxidative DNA damage and RNA turnover in the body.

5.5.3.5 Glycolysis

Lactate, an important end product of glycolysis, was found to be elevated in BC tissue and urine [115, 124], indicating an increased rate of glycolysis rate. The upregulation of glycolysis, resulting in increased glucose consumption, is a universal phenomenon in cancer and is termed the "Warburg effect" [125, 126]. Gatenby and Gillies proposed that the upregulation of glycolysis is an adaptation of premalignant lesions to intermittent hypoxia, but requires evolution to the resultant proliferative and invasive phenotypes where resistance to acid-induced cell toxicity is also observed [125].

Diagnosis and prognosis of various diseases are enhanced by the identification of biomarkers, which can differentiate individuals with the disease from those without. Ideal markers are easily detectable in tissue, serum, and urine, and have a high sensitivity and specificity. There are several potential applications of metabolomics in BC and other cancers; this includes improving detection, providing prognostic information, and impacting treatment.

5.6 Clinically Applicable BC Biomarkers-Based Tools

At present, the FDA has approved six tests for detecting or monitoringBC. NMP22, NMP22 BladderChek, and UroVysion have FDA approval for BC diagnosis and surveillance; Immunocytology (uCyt+), BTA-TRAK, and BTA-STAT have been approved only for surveillance [127–131]. There are also many metabolites that can be considered as potential tumor biomarkers for BC.

By ultra-performance liquid chromatography time-of-flight mass spectrometry, imidazole-acetic acid was evidenced in BC [132]. A metabolite panel consisting of indolylacryloylglycine, N2-galacturonyl-L-lysine, and aspartylglutamate can discriminate high- vs. low-grade BC [133]. In addition, alterations in the metabolisms of phenylalanine, arginine, proline, and tryptophan were evidenced by UPLC-MS in NMBIC [134]. Jin X et al. confirmed through their study that carnitine acyltransferase and pyruvate dehydrogenase complex expressions are significantly altered in cancer [41]. Alberice JV et al. propose that metabolites related to the tryptophan metabolism pathway, such as kynurenine and tryptophan, are potential urinary biomarkers and therapeutic targets of BC therapy [116]. Wittmann et al. performed unbiased metabolomics on a set of urine samples from BC patients, revealing nearly 1000 distinct metabolic signatures, of which 587 have a chemical identity [135]. The authors chose a set of 25 potential biomarkers from this group and tested this panel on a second independent cohort to validate its predictive power. A new group of metabolites, including lactate, adenosine, succinate, and palmitoyl sphingomyelin, were proposed as urinary biomarkers; thus, showing the involvement of lipid metabolism in BC progression.

5.7 Conclusions and Perspectives

At present, there is much research on biomarkers of BC. Biomarkers can be identified in tissue, blood, urine, etc. and include genes, proteins, metabolites, etc. In this paper, we summarized the research progress of BC biomarkers in recent years. Due to the advantages of urine collection, including non-invasive procedures, simplicity, easy storage, low-cost, and direct contact with bladder cancer tissue, we focused particularly on urinary biomarker research progress. Compared to genomics and proteomics, metabonomics of BC is still in its early stages. However, because of the great progress in metabonomics research in BC using NMR, GC-MS, and LC-MS, metabonomics has been widely used to propose new biomarkers. These may be applied to screening, diagnosing, treating, evaluating, and monitoring BC. Although the potential of metabonomics to improve detection and treatment of BC may be great, the main limitation is the lack of reliable validation for a large population. Current research has so far been limited to smaller samples without validation and metabolites can be easily affected by various factors. For future metabonomics research, experimental design and analysis methods need to be standardized to eliminate the systemic influence of confounding variables on the measurement of metabolites, make results more comparable, verify potential biomarkers, and assist in clinical applications against BC.

References

1. Siegel, R.L., Miller, K.D., Jemal, A.: Cancer statistics, 2019. CA Cancer J. Clin. **69**, 7–34 (2019)
2. Mitra, A.P., Cote, R.J.: Molecular pathogenesis and diagnostics of bladder cancer. Annu. Rev. Pathol. **4**, 251–285 (2009)
3. Ruder, A.M., Fine, L.J., Sundin, D.S.: National estimates of occupational exposure to animal bladder tumorigens. J. Occup. Med. **32**, 797–805 (1990)
4. Pelucchi, C., Bosetti, C., Negri, E., Malvezzi, M., La Vecchia, C.: Mechanisms of disease: the epidemiology of bladder cancer. Nat. Clin. Pract. Urol. **3**, 327–340 (2006)
5. Castelao, J.E., Yuan, J.M., Gago-Dominguez, M., Yu, M.C., Ross, R.K.: Non-steroidal anti-inflammatory drugs and bladder cancer prevention. Br. J. Cancer. **82**, 1364–1369 (2000)
6. Lokeshwar, V.B., Habuchi, T., Grossman, H.B., Murphy, W.M., Hautmann, S.H., Hemstreet 3rd, G.P., Bono, A.V., Getzenberg, R.H., Goebell, P., Schmitz-Drager, B.J., et al.: Bladder tumor markers beyond cytology: international consensus panel on bladder tumor markers. Urology. **66**, 35–63 (2005)
7. Sloan, F.A., Yashkin, A.P., Akushevich, I., Inman, B.A.: The cost to medicare of bladder cancer care. Eur Urol Oncol. **3**, 515–522 (2019)
8. James, A.C., Gore, J.L.: The costs of non-muscle invasive bladder cancer. Urol. Clin. North Am. **40**, 261–269 (2013)
9. Moch, H., Cubilla, A.L., Humphrey, P.A., Reuter, V.E., Ulbright, T.M.: The 2016 WHO classification of tumours of the urinary system and male genital organs-part a: renal, penile, and testicular tumours. Eur. Urol. **70**, 93–105 (2016)
10. Eich, M.L., Dyrskjot, L., Netto, G.J.: Toward personalized management in bladder cancer: the promise of novel molecular taxonomy. Virchows Arch. **471**, 271–280 (2017)
11. Kirkali, Z., Chan, T., Manoharan, M., Algaba, F., Busch, C., Cheng, L., Kiemeney, L., Kriegmair, M., Montironi, R., Murphy, W.M., et al.: Bladder cancer: epidemiology, staging and grading, and diagnosis. Urology. **66**, 4–34 (2005)

12. Wang, G., McKenney, J.K.: Urinary bladder pathology: World Health Organization classification and American joint committee on cancer staging update. Arch. Pathol. Lab. Med. **143**, 571–577 (2019)
13. Reuter, V.E.: The pathology of bladder cancer. Urology. **67**, 11–17 (2006) discussion 17–18
14. Damrauer, J.S., Hoadley, K.A., Chism, D.D., Fan, C., Tiganelli, C.J., Wobker, S.E., Yeh, J.J., Milowsky, M.I., Iyer, G., Parker, J.S., et al.: Intrinsic subtypes of high-grade bladder cancer reflect the hallmarks of breast cancer biology. Proc. Natl. Acad. Sci. U. S. A. **111**, 3110–3115 (2014)
15. Choi, W., Porten, S., Kim, S., Willis, D., Plimack, E.R., Hoffman-Censits, J., Roth, B., Cheng, T., Tran, M., Lee, I.L., et al.: Identification of distinct basal and luminal subtypes of muscle-invasive bladder cancer with different sensitivities to frontline chemotherapy. Cancer Cell. **25**, 152–165 (2014)
16. Cancer Genome Atlas Research, N.: Comprehensive molecular characterization of urothelial bladder carcinoma. Nature. **507**, 315–322 (2014)
17. Sjodahl, G., Lauss, M., Lovgren, K., Chebil, G., Gudjonsson, S., Veerla, S., Patschan, O., Aine, M., Ferno, M., Ringner, M., et al.: A molecular taxonomy for urothelial carcinoma. Clin. Cancer Res. **18**, 3377–3386 (2012)
18. Kim, J., Akbani, R., Creighton, C.J., Lerner, S.P., Weinstein, J.N., Getz, G., Kwiatkowski, D.J.: Invasive bladder cancer: genomic insights and therapeutic promise. Clin. Cancer Res. **21**, 4514–4524 (2015)
19. Sjodahl, G., Eriksson, P., Liedberg, F., Hoglund, M.: Molecular classification of urothelial carcinoma: global mRNA classification versus tumour-cell phenotype classification. J. Pathol. **242**, 113–125 (2017)
20. Kardos, J., Chai, S., Mose, L.E., Selitsky, S.R., Krishnan, B., Saito, R., Iglesia, M.D., Milowsky, M.I., Parker, J.S., Kim, W.Y., et al.: Claudin-low bladder tumors are immune infiltrated and actively immune suppressed. JCI Insight. **1**, e85902 (2016)
21. Hurle, R., Losa, A., Manzetti, A., Lembo, A.: Upper urinary tract tumors developing after treatment of superficial bladder cancer: 7-year follow-up of 591 consecutive patients. Urology. **53**, 1144–1148 (1999)
22. Lotan, Y., Svatek, R.S., Sagalowsky, A.I.: Should we screen for bladder cancer in a high-risk population?: a cost per life-year saved analysis. Cancer. **107**, 982–990 (2006)
23. Daneshmand, S., Bazargani, S.T., Bivalacqua, T.J., Holzbeierlein, J.M., Willard, B., Taylor, J.M., Liao, J.C., Pohar, K., Tierney, J., Konety, B., et al.: Blue light cystoscopy for the diagnosis of bladder cancer: results from the US prospective multicenter registry. Urol Oncol. **36**, 361.e1–361.e6 (2018)
24. de Martino, M., Shariat, S.F., Hofbauer, S.L., Lucca, I., Taus, C., Wiener, H.G., Haitel, A., Susani, M., Klatte, T.: Aurora A kinase as a diagnostic urinary marker for urothelial bladder cancer. World J. Urol. **33**, 105–110 (2015)
25. Arnold Egloff, S.A., Du, L., Loomans, H.A., Starchenko, A., Su, P.F., Ketova, T., Knoll, P.B., Wang, J., Haddad, A.Q., Fadare, O., et al.: Shed urinary ALCAM is an independent prognostic biomarker of three-year overall survival after cystectomy in patients with bladder cancer. Oncotarget. **8**, 722–741 (2017)
26. Pozzi, V., Di Ruscio, G., Sartini, D., Campagna, R., Seta, R., Fulvi, P., Vici, A., Milanese, G., Brandoni, G., Galosi, A.B., et al.: Clinical performance and utility of a NNMT-based urine test for bladder cancer. Int. J. Biol. Markers. **33**, 94–101 (2018)
27. Choi, S., Shin, J.H., Lee, Y.R., Joo, H.K., Song, K.H., Na, Y.G., Chang, S.J., Lim, J.S., Jeon, B.H.: Urinary APE1/Ref-1: a potential bladder cancer biomarker. Dis. Markers. **2016**, 7276502 (2016)
28. Mi, Y., Zhao, Y., Shi, F., Zhang, M., Wang, C., Liu, X.: Diagnostic accuracy of urine cytokeratin 20 for bladder cancer: a meta-analysis. Asia Pac. J. Clin. Oncol. **15**, e11–e19 (2019)
29. Ecke, T.H., Weiss, S., Stephan, C., Hallmann, S., Barski, D., Otto, T., Gerullis, H.: UBC((R)) rapid test for detection of carcinoma in situ for bladder cancer. Tumour Biol. **39**, 1010428317701624 (2017)
30. Chen, Y.T., Chen, C.L., Chen, H.W., Chung, T., Wu, C.C., Chen, C.D., Hsu, C.W., Chen, M.C., Tsui, K.H., Chang, P.L., et al.: Discovery of novel bladder cancer biomarkers by comparative urine proteomics using iTRAQ technology. J. Proteome Res. **9**, 5803–5815 (2010)
31. Chen, Y.T., Chen, H.W., Domanski, D., Smith, D.S., Liang, K.H., Wu, C.C., Chen, C.L., Chung, T., Chen, M.C., Chang, Y.S., et al.: Multiplexed quantification of 63 proteins in human urine by multiple reaction monitoring-based mass spectrometry for discovery of potential bladder cancer biomarkers. J. Proteome. **75**, 3529–3545 (2012)
32. Schiffer, E., Vlahou, A., Petrolekas, A., Stravodimos, K., Tauber, R., Geschwend, J.E., Neuhaus, J., Stolzenburg, J.U., Conaway, M.R., Mischak, H., et al.: Prediction of muscle-invasive bladder cancer using urinary proteomics. Clin. Cancer Res. **15**, 4935–4943 (2009)
33. Masuda, N., Ogawa, O., Park, M., Liu, A.Y., Goodison, S., Dai, Y., Kozai, L., Furuya, H., Lotan, Y., Rosser, C.J., et al.: Meta-analysis of a 10-plex urine-based biomarker assay for the detection of bladder cancer. Oncotarget. **9**, 7101–7111 (2018)
34. Soukup, V., Kalousova, M., Capoun, O., Sobotka, R., Breyl, Z., Pesl, M., Zima, T., Hanus, T.: Panel of urinary diagnostic markers for non-invasive detection of primary and recurrent urothelial urinary bladder carcinoma. Urol. Int. **95**, 56–64 (2015)
35. Shabayek, M.I., Sayed, O.M., Attaia, H.A., Awida, H.A., Abozeed, H.: Diagnostic evaluation of urinary angiogenin (ANG) and clusterin (CLU) as biomarker for bladder cancer. Pathol. Oncol. Res. **20**, 859–866 (2014)
36. Salomo, K., Huebner, D., Boehme, M.U., Herr, A., Brabetz, W., Heberling, U., Hakenberg, O.W., Jahn,

D., Grimm, M.O., Steinbach, D., et al.: Urinary transcript quantitation of CK20 and IGF2 for the non-invasive bladder cancer detection. J. Cancer Res. Clin. Oncol. **143**, 1757–1769 (2017)
37. Snell, K.I.E., Ward, D.G., Gordon, N.S., Goldsmith, J.C., Sutton, A.J., Patel, P., James, N.D., Zeegers, M.P., Cheng, K.K., Bryan, R.T.: Exploring the roles of urinary HAI-1, EpCAM & EGFR in bladder cancer prognosis & risk stratification. Oncotarget. **9**, 25244–25253 (2018)
38. Yang, Y., Xu, J., Zhang, Q.: Detection of urinary survivin using a magnetic particles-based chemiluminescence immunoassay for the preliminary diagnosis of bladder cancer and renal cell carcinoma combined with LAPTM4B. Oncol. Lett. **15**, 7923–7933 (2018)
39. Santi, R., Cai, T., Nobili, S., Galli, I.C., Amorosi, A., Comperat, E., Nesi, G.: Snail immunohistochemical overexpression correlates to recurrence risk in non-muscle invasive bladder cancer: results from a longitudinal cohort study. Virchows Arch. **472**, 605–613 (2018)
40. Azevedo, R., Soares, J., Gaiteiro, C., Peixoto, A., Lima, L., Ferreira, D., Relvas-Santos, M., Fernandes, E., Tavares, A., Cotton, S., et al.: Glycan affinity magnetic nanoplatforms for urinary glycobiomarkers discovery in bladder cancer. Talanta. **184**, 347–355 (2018)
41. Jin, X., Yun, S.J., Jeong, P., Kim, I.Y., Kim, W.J., Park, S.: Diagnosis of bladder cancer and prediction of survival by urinary metabolomics. Oncotarget. **5**, 1635–1645 (2014)
42. Zhou, Y., Song, R., Ma, C., Zhou, L., Liu, X., Yin, P., Zhang, Z., Sun, Y., Xu, C., Lu, X., et al.: Discovery and validation of potential urinary biomarkers for bladder cancer diagnosis using a pseudotargeted GC-MS metabolomics method. Oncotarget. **8**, 20719–20728 (2017)
43. Cheng, X., Liu, X., Liu, X., Guo, Z., Sun, H., Zhang, M., Ji, Z., Sun, W.: Metabolomics of non-muscle invasive bladder cancer: biomarkers for early detection of bladder cancer. Front. Oncol. **8**, 494 (2018)
44. Tan, Y., Yin, P., Tang, L., Xing, W., Huang, Q., Cao, D., Zhao, X., Wang, W., Lu, X., Xu, Z., et al.: Metabolomics study of stepwise hepatocarcinogenesis from the model rats to patients: potential biomarkers effective for small hepatocellular carcinoma diagnosis. Mol. Cell. Proteomics. **11**(M111), 010694 (2012)
45. Yumba Mpanga, A., Siluk, D., Jacyna, J., Szerkus, O., Wawrzyniak, R., Markuszewski, M., Matuszewski, M., Kaliszan, R., Markuszewski, M.J.: Targeted metabolomics in bladder cancer: From analytical methods development and validation towards application to clinical samples. Anal. Chim. Acta. **1037**, 188–199 (2018)
46. Shen, C., Sun, Z., Chen, D., Su, X., Jiang, J., Li, G., Lin, B., Yan, J.: Developing urinary metabolomic signatures as early bladder cancer diagnostic markers. OMICS. **19**, 1–11 (2015)
47. Su, S.F., de Castro Abreu, A.L., Chihara, Y., Tsai, Y., Andreu-Vieyra, C., Daneshmand, S., Skinner, E.C., Jones, P.A., Siegmund, K.D., Liang, G.: A panel of three markers hyper- and hypomethylated in urine sediments accurately predicts bladder cancer recurrence. Clin. Cancer Res. **20**, 1978–1989 (2014)
48. Pu, R.T., Laitala, L.E., Clark, D.P.: Methylation profiling of urothelial carcinoma in bladder biopsy and urine. Acta Cytol. **50**, 499–506 (2006)
49. Hauser, S., Kogej, M., Fechner, G., Von Pezold, J., Vorreuther, R., Lummen, G., Muller, S.C., Ellinger, J.: Serum DNA hypermethylation in patients with bladder cancer: results of a prospective multicenter study. Anticancer Res. **33**, 779–784 (2013)
50. Renard, I., Joniau, S., van Cleynenbreugel, B., Collette, C., Naome, C., Vlassenbroeck, I., Nicolas, H., de Leval, J., Straub, J., Van Criekinge, W., et al.: Identification and validation of the methylated TWIST1 and NID2 genes through real-time methylation-specific polymerase chain reaction assays for the noninvasive detection of primary bladder cancer in urine samples. Eur. Urol. **58**, 96–104 (2010)
51. Fantony, J.J., Longo, T.A., Gopalakrishna, A., Owusu, R., Lance, R.S., Foo, W.C., Inman, B.A., Abern, M.R.: Urinary NID2 and TWIST1 methylation to augment conventional urine cytology for the detection of bladder cancer. Cancer Biomark. **18**, 381–387 (2017)
52. van der Heijden, A.G., Mengual, L., Ingelmo-Torres, M., Lozano, J.J., van Rijt-van de Westerlo, C.C.M., Baixauli, M., Geavlete, B., Moldoveanud, C., Ene, C., Dinney, C.P., et al.: Urine cell-based DNA methylation classifier for monitoring bladder cancer. Clin. Epigenetics. **10**, 71 (2018)
53. Wang, Y., Yu, Y., Ye, R., Zhang, D., Li, Q., An, D., Fang, L., Lin, Y., Hou, Y., Xu, A., et al.: An epigenetic biomarker combination of PCDH17 and POU4F2 detects bladder cancer accurately by methylation analyses of urine sediment DNA in Han Chinese. Oncotarget. **7**, 2754–2764 (2016)
54. Roperch, J.P., Grandchamp, B., Desgrandchamps, F., Mongiat-Artus, P., Ravery, V., Ouzaid, I., Roupret, M., Phe, V., Ciofu, C., Tubach, F., et al.: Promoter hypermethylation of HS3ST2, SEPTIN9 and SLIT2 combined with FGFR3 mutations as a sensitive/specific urinary assay for diagnosis and surveillance in patients with low or high-risk non-muscle-invasive bladder cancer. BMC Cancer. **16**, 704 (2016)
55. Pietrusinski, M., Kepczynski, J.A., Borkowska, E., Traczyk-Borszynska, M., Constantinou, M., Kauzewski, B., Borowiec, M.: Detection of bladder cancer in urine sediments by a hypermethylation panel of selected tumor suppressor genes. Cancer Biomark. **18**, 47–59 (2017)
56. Sasaki, H., Yoshiike, M., Nozawa, S., Usuba, W., Katsuoka, Y., Aida, K., Kitajima, K., Kudo, H., Hoshikawa, M., Yoshioka, Y., et al.: Expression level of urinary microRNA-146a-5p is increased in patients with bladder cancer and decreased in those after transurethral resection. Clin. Genitourin. Cancer. **14**, e493–e499 (2016)
57. Hanke, M., Hoefig, K., Merz, H., Feller, A.C., Kausch, I., Jocham, D., Warnecke, J.M., Sczakiel, G.:

A robust methodology to study urine microRNA as tumor marker: microRNA-126 and microRNA-182 are related to urinary bladder cancer. Urol. Oncol. **28**, 655–661 (2010)
58. Wiklund, E.D., Gao, S., Hulf, T., Sibbritt, T., Nair, S., Costea, D.E., Villadsen, S.B., Bakholdt, V., Bramsen, J.B., Sorensen, J.A., et al.: MicroRNA alterations and associated aberrant DNA methylation patterns across multiple sample types in oral squamous cell carcinoma. PLoS One. **6**, e27840 (2011)
59. Chen, Y.H., Wang, S.Q., Wu, X.L., Shen, M., Chen, Z.G., Chen, X.G., Liu, Y.X., Zhu, X.L., Guo, F., Duan, X.Z., et al.: Characterization of microRNAs expression profiling in one group of Chinese urothelial cell carcinoma identified by Solexa sequencing. Urol. Oncol. **31**, 219–227 (2013)
60. Eissa, S., Matboli, M., Essawy, N.O., Kotb, Y.M.: Integrative functional genetic-epigenetic approach for selecting genes as urine biomarkers for bladder cancer diagnosis. Tumour Biol. **36**, 9545–9552 (2015)
61. Pospisilova, S., Pazourkova, E., Horinek, A., Brisuda, A., Svobodova, I., Soukup, V., Hrbacek, J., Capoun, O., Hanus, T., Mares, J., et al.: MicroRNAs in urine supernatant as potential non-invasive markers for bladder cancer detection. Neoplasma. **63**, 799–808 (2016)
62. Pignot, G., Cizeron-Clairac, G., Vacher, S., Susini, A., Tozlu, S., Vieillefond, A., Zerbib, M., Lidereau, R., Debre, B., Amsellem-Ouazana, D., et al.: microRNA expression profile in a large series of bladder tumors: identification of a 3-miRNA signature associated with aggressiveness of muscle-invasive bladder cancer. Int. J. Cancer. **132**, 2479–2491 (2013)
63. Yun, S.J., Jeong, P., Kim, W.T., Kim, T.H., Lee, Y.S., Song, P.H., Choi, Y.H., Kim, I.Y., Moon, S.K., Kim, W.J.: Cell-free microRNAs in urine as diagnostic and prognostic biomarkers of bladder cancer. Int. J. Oncol. **41**, 1871–1878 (2012)
64. Matsushita, R., Seki, N., Chiyomaru, T., Inoguchi, S., Ishihara, T., Goto, Y., Nishikawa, R., Mataki, H., Tatarano, S., Itesako, T., et al.: Tumour-suppressive microRNA-144-5p directly targets CCNE1/2 as potential prognostic markers in bladder cancer. Br. J. Cancer. **113**, 282–289 (2015)
65. Zhang, D.Z., Lau, K.M., Chan, E.S., Wang, G., Szeto, C.C., Wong, K., Choy, R.K., Ng, C.F.: Cell-free urinary microRNA-99a and microRNA-125b are diagnostic markers for the non-invasive screening of bladder cancer. PLoS One. **9**, e100793 (2014)
66. Tolle, A., Jung, M., Rabenhorst, S., Kilic, E., Jung, K., Weikert, S.: Identification of microRNAs in blood and urine as tumour markers for the detection of urinary bladder cancer. Oncol. Rep. **30**, 1949–1956 (2013)
67. Juracek, J., Peltanova, B., Dolezel, J., Fedorko, M., Pacik, D., Radova, L., Vesela, P., Svoboda, M., Slaby, O., Stanik, M.: Genome-wide identification of urinary cell-free microRNAs for non-invasive detection of bladder cancer. J. Cell. Mol. Med. **22**, 2033–2038 (2018)
68. Bathe, O.F., Farshidfar, F.: From genotype to functional phenotype: unraveling the metabolomic features of colorectal cancer. Genes (Basel). **5**, 536–560 (2014)
69. Dinges, S.S., Hohm, A., Vandergrift, L.A., Nowak, J., Habbel, P., Kaltashov, I.A., Cheng, L.L.: Cancer metabolomic markers in urine: evidence, techniques and recommendations. Nat. Rev. Urol. **16**, 339–362 (2019)
70. Roberts, L.D., Souza, A.L., Gerszten, R.E., Clish, C.B.: Targeted metabolomics. Curr Protoc Mol Biol. *Chapter 30*, Unit 30. **32**, 31–24 (2012)
71. Alonso, A., Marsal, S., Julia, A.: Analytical methods in untargeted metabolomics: state of the art in 2015. Front. Bioeng. Biotechnol. **3**, 23 (2015)
72. Weiss, R.H., Kim, K.: Metabolomics in the study of kidney diseases. Nat. Rev. Nephrol. **8**, 22–33 (2011)
73. Emwas, A.H.: The strengths and weaknesses of NMR spectroscopy and mass spectrometry with particular focus on metabolomics research. Methods Mol. Biol. **1277**, 161–193 (2015)
74. Kalim, S., Rhee, E.P.: An overview of renal metabolomics. Kidney Int. **91**, 61–69 (2017)
75. Griffin, J.L., Shockcor, J.P.: Metabolic profiles of cancer cells. Nat. Rev. Cancer. **4**, 551–561 (2004)
76. Cheng, Y., Yang, X., Deng, X., Zhang, X., Li, P., Tao, J., Qin, C., Wei, J., Lu, Q.: Metabolomics in bladder cancer: a systematic review. Int. J. Clin. Exp. Med. **8**, 11052–11063 (2015)
77. Assfalg, M., Bertini, I., Colangiuli, D., Luchinat, C., Schafer, H., Schutz, B., Spraul, M.: Evidence of different metabolic phenotypes in humans. Proc. Natl. Acad. Sci. U. S. A. **105**, 1420–1424 (2008)
78. Lotan, Y., Shariat, S.F., Schmitz-Drager, B.J., Sanchez-Carbayo, M., Jankevicius, F., Racioppi, M., Minner, S.J., Stohr, B., Bassi, P.F., Grossman, H.B.: Considerations on implementing diagnostic markers into clinical decision making in bladder cancer. Urol. Oncol. **28**, 441–448 (2010)
79. Barderas, M.G., Laborde, C.M., Posada, M., de la Cuesta, F., Zubiri, I., Vivanco, F., Alvarez-Llamas, G.: Metabolomic profiling for identification of novel potential biomarkers in cardiovascular diseases. J. Biomed. Biotechnol. **2011**, 790132 (2011)
80. Beckonert, O., Keun, H.C., Ebbels, T.M., Bundy, J., Holmes, E., Lindon, J.C., Nicholson, J.K.: Metabolic profiling, metabolomic and metabonomic procedures for NMR spectroscopy of urine, plasma, serum and tissue extracts. Nat. Protoc. **2**, 2692–2703 (2007)
81. Ellis, D.I., Dunn, W.B., Griffin, J.L., Allwood, J.W., Goodacre, R.: Metabolic fingerprinting as a diagnostic tool. Pharmacogenomics. **8**, 1243–1266 (2007)
82. Mapelli, V., Olsson, L., Nielsen, J.: Metabolic footprinting in microbiology: methods and applications in functional genomics and biotechnology. Trends Biotechnol. **26**, 490–497 (2008)
83. Lindon, J.C., Nicholson, J.K.: Spectroscopic and statistical techniques for information recovery in metabonomics and metabolomics. Annu Rev Anal Chem (Palo Alto, Calif). **1**, 45–69 (2008)

84. Kaddurah-Daouk, R., Kristal, B.S., Weinshilboum, R.M.: Metabolomics: a global biochemical approach to drug response and disease. Annu. Rev. Pharmacol. Toxicol. **48**, 653–683 (2008)
85. Fan, X., Bai, J., Shen, P.: Diagnosis of breast cancer using HPLC metabonomics fingerprints coupled with computational methods. Conf. Proc. IEEE Eng. Med. Biol. Soc. **6**, 6081–6084 (2005)
86. van Doorn, M., Vogels, J., Tas, A., van Hoogdalem, E.J., Burggraaf, J., Cohen, A., van der Greef, J.: Evaluation of metabolite profiles as biomarkers for the pharmacological effects of thiazolidinediones in Type 2 diabetes mellitus patients and healthy volunteers. Br. J. Clin. Pharmacol. **63**, 562–574 (2007)
87. Clayton, T.A., Lindon, J.C., Cloarec, O., Antti, H., Charuel, C., Hanton, G., Provost, J.P., Le Net, J.L., Baker, D., Walley, R.J., et al.: Pharmaco-metabonomic phenotyping and personalized drug treatment. Nature. **440**, 1073–1077 (2006)
88. Portilla, D., Li, S., Nagothu, K.K., Megyesi, J., Kaissling, B., Schnackenberg, L., Safirstein, R.L., Beger, R.D.: Metabolomic study of cisplatin-induced nephrotoxicity. Kidney Int. **69**, 2194–2204 (2006)
89. Fiehn, O.: Metabolomics – the link between genotypes and phenotypes. Plant Mol. Biol. **48**, 155–171 (2002)
90. Kind, T., Wohlgemuth, G., Lee, D.Y., Lu, Y., Palazoglu, M., Shahbaz, S., Fiehn, O.: FiehnLib: mass spectral and retention index libraries for metabolomics based on quadrupole and time-of-flight gas chromatography/mass spectrometry. Anal. Chem. **81**, 10038–10048 (2009)
91. Becker, S., Kortz, L., Helmschrodt, C., Thiery, J., Ceglarek, U.: LC-MS-based metabolomics in the clinical laboratory. J. Chromatogr. B Analyt. Technol. Biomed. Life Sci. **883-884**, 68–75 (2012)
92. Gika, H.G., Theodoridis, G.A., Plumb, R.S., Wilson, I.D.: Current practice of liquid chromatography-mass spectrometry in metabolomics and metabonomics. J. Pharm. Biomed. Anal. **87**, 12–25 (2014)
93. Dunn, W.B., Bailey, N.J., Johnson, H.E.: Measuring the metabolome: current analytical technologies. Analyst. **130**, 606–625 (2005)
94. Zhang, A., Sun, H., Wang, P., Han, Y., Wang, X.: Modern analytical techniques in metabolomics analysis. Analyst. **137**, 293–300 (2012)
95. Myint, K.T., Uehara, T., Aoshima, K., Oda, Y.: Polar anionic metabolome analysis by nano-LC/MS with a metal chelating agent. Anal. Chem. **81**, 7766–7772 (2009)
96. Fernandez-Arroyo, S., Gomez-Martinez, A., Rocamora-Reverte, L., Quirantes-Pine, R., Segura-Carretero, A., Fernandez-Gutierrez, A., Ferragut, J.A.: Application of nanoLC-ESI-TOF-MS for the metabolomic analysis of phenolic compounds from extra-virgin olive oil in treated colon-cancer cells. J. Pharm. Biomed. Anal. **63**, 128–134 (2012)
97. Ramautar, R., Somsen, G.W., de Jong, G.J.: CE-MS for metabolomics: developments and applications in the period 2012–2014. Electrophoresis. **36**, 212–224 (2015)
98. Naz, S., Garcia, A., Rusak, M., Barbas, C.: Method development and validation for rat serum fingerprinting with CE-MS: application to ventilator-induced-lung-injury study. Anal. Bioanal. Chem. **405**, 4849–4858 (2013)
99. Moraes, E.P., Ruperez, F.J., Plaza, M., Herrero, M., Barbas, C.: Metabolomic assessment with CE-MS of the nutraceutical effect of Cystoseira spp extracts in an animal model. Electrophoresis. **32**, 2055–2062 (2011)
100. Shi, H., Vigneau-Callahan, K.E., Matson, W.R., Kristal, B.S.: Attention to relative response across sequential electrodes improves quantitation of coulometric array. Anal. Biochem. **302**, 239–245 (2002)
101. Shi, H., Paolucci, U., Vigneau-Callahan, K.E., Milbury, P.E., Matson, W.R., Kristal, B.S.: Development of biomarkers based on diet-dependent metabolic serotypes: practical issues in development of expert system-based classification models in metabolomic studies. OMICS. **8**, 197–208 (2004)
102. Paolucci, U., Vigneau-Callahan, K.E., Shi, H., Matson, W.R., Kristal, B.S.: Development of biomarkers based on diet-dependent metabolic serotypes: concerns and approaches for cohort and gender issues in serum metabolome studies. OMICS. **8**, 209–220 (2004b)
103. Paolucci, U., Vigneau-Callahan, K.E., Shi, H., Matson, W.R., Kristal, B.S.: Development of biomarkers based on diet-dependent metabolic serotypes: characteristics of component-based models of metabolic serotypes. OMICS. **8**, 221–238 (2004a)
104. Beal, M.F., Matson, W.R., Storey, E., Milbury, P., Ryan, E.A., Ogawa, T., Bird, E.D.: Kynurenic acid concentrations are reduced in Huntington's disease cerebral cortex. J. Neurol. Sci. **108**, 80–87 (1992)
105. Massari, F., Ciccarese, C., Santoni, M., Iacovelli, R., Mazzucchelli, R., Piva, F., Scarpelli, M., Berardi, R., Tortora, G., Lopez-Beltran, A., et al.: Metabolic phenotype of bladder cancer. Cancer Treat. Rev. **45**, 46–57 (2016)
106. Sahu, D., Lotan, Y., Wittmann, B., Neri, B., Hansel, D.E.: Metabolomics analysis reveals distinct profiles of nonmuscle-invasive and muscle-invasive bladder cancer. Cancer Med. **6**, 2106–2120 (2017)
107. Srivastava, S., Roy, R., Singh, S., Kumar, P., Dalela, D., Sankhwar, S.N., Goel, A., Sonkar, A.A.: Taurine – a possible fingerprint biomarker in non-muscle invasive bladder cancer: a pilot study by 1H NMR spectroscopy. Cancer Biomark. **6**, 11–20 (2010)
108. Anghileri, L.J., Crone-Escanye, M.C., Thouvenot, P., Brunotte, F., Robert, J.: Mechanisms of gallium-67 accumulation by tumors: role of cell membrane permeability. J. Nucl. Med. **29**, 663–668 (1988)
109. Mycielska, M.E., Patel, A., Rizaner, N., Mazurek, M.P., Keun, H., Patel, A., Ganapathy, V., Djamgoz, M.B.: Citrate transport and metabolism in mammalian cells: prostate epithelial cells and prostate cancer. BioEssays. **31**, 10–20 (2009)
110. Huang, Z., Lin, L., Gao, Y., Chen, Y., Yan, X., Xing, J., Hang, W.: Bladder cancer determination via two

urinary metabolites: a biomarker pattern approach. Mol. Cell. Proteomics. **10**(M111), 007922 (2011)
111. Putluri, N., Shojaie, A., Vasu, V.T., Vareed, S.K., Nalluri, S., Putluri, V., Thangjam, G.S., Panzitt, K., Tallman, C.T., Butler, C., et al.: Metabolomic profiling reveals potential markers and bioprocesses altered in bladder cancer progression. Cancer Res. **71**, 7376–7386 (2011)
112. Peluso, G., Nicolai, R., Reda, E., Benatti, P., Barbarisi, A., Calvani, M.: Cancer and anticancer therapy-induced modifications on metabolism mediated by carnitine system. J. Cell. Physiol. **182**, 339–350 (2000)
113. Pelicano, H., Carney, D., Huang, P.: ROS stress in cancer cells and therapeutic implications. Drug Resist. Updat. **7**, 97–110 (2004)
114. Shen, H., Kauvar, L., Tew, K.D.: Importance of glutathione and associated enzymes in drug response. Oncol. Res. **9**, 295–302 (1997)
115. Pasikanti, K.K., Esuvaranathan, K., Hong, Y., Ho, P.C., Mahendran, R., Raman Nee Mani, L., Chiong, E., Chan, E.C.: Urinary metabotyping of bladder cancer using two-dimensional gas chromatography time-of-flight mass spectrometry. J. Proteome Res. **12**, 3865–3873 (2013)
116. Alberice, J.V., Amaral, A.F., Armitage, E.G., Lorente, J.A., Algaba, F., Carrilho, E., Marquez, M., Garcia, A., Malats, N., Barbas, C.: Searching for urine biomarkers of bladder cancer recurrence using a liquid chromatography-mass spectrometry and capillary electrophoresis-mass spectrometry metabolomics approach. J. Chromatogr. A. **1318**, 163–170 (2013)
117. Bansal, N., Gupta, A., Mitash, N., Shakya, P.S., Mandhani, A., Mahdi, A.A., Sankhwar, S.N., Mandal, S.K.: Low- and high-grade bladder cancer determination via human serum-based metabolomics approach. J. Proteome Res. **12**, 5839–5850 (2013)
118. Chung, K.T., Gadupudi, G.S.: Possible roles of excess tryptophan metabolites in cancer. Environ. Mol. Mutagen. **52**, 81–104 (2011)
119. Opitz, C.A., Litzenburger, U.M., Sahm, F., Ott, M., Tritschler, I., Trump, S., Schumacher, T., Jestaedt, L., Schrenk, D., Weller, M., et al.: An endogenous tumour-promoting ligand of the human aryl hydrocarbon receptor. Nature. **478**, 197–203 (2011)
120. Rosado, J.O., Salvador, M., Bonatto, D.: Importance of the trans-sulfuration pathway in cancer prevention and promotion. Mol. Cell. Biochem. **301**, 1–12 (2007)
121. Zheng, Y.F., Kong, H.W., Xiong, J.H., Lv, S., Xu, G.W.: Clinical significance and prognostic value of urinary nucleosides in breast cancer patients. Clin. Biochem. **38**, 24–30 (2005)
122. Wu, H., Xue, R., Dong, L., Liu, T., Deng, C., Zeng, H., Shen, X.: Metabolomic profiling of human urine in hepatocellular carcinoma patients using gas chromatography/mass spectrometry. Anal. Chim. Acta. **648**, 98–104 (2009)
123. Nakano, K., Nakao, T., Schram, K.H., Hammargren, W.M., McClure, T.D., Katz, M., Petersen, E.: Urinary excretion of modified nucleosides as biological marker of RNA turnover in patients with cancer and AIDS. Clin. Chim. Acta. **218**, 169–183 (1993)
124. Tripathi, P., Somashekar, B.S., Ponnusamy, M., Gursky, A., Dailey, S., Kunju, P., Lee, C.T., Chinnaiyan, A.M., Rajendiran, T.M., Ramamoorthy, A.: HR-MAS NMR tissue metabolomic signatures cross-validated by mass spectrometry distinguish bladder cancer from benign disease. J. Proteome Res. **12**, 3519–3528 (2013)
125. Gatenby, R.A., Gillies, R.J.: Why do cancers have high aerobic glycolysis? Nat. Rev. Cancer. **4**, 891–899 (2004)
126. Semenza, G.L., Artemov, D., Bedi, A., Bhujwalla, Z., Chiles, K., Feldser, D., Laughner, E., Ravi, R., Simons, J., Taghavi, P., et al.: 'The metabolism of tumours': 70 years later. Novartis Found. Symp. **240**, 251–260 (2001) discussion 260-254
127. Liang, Q., Zhang, G., Li, W., Wang, J., Sheng, S.: Comparison of the diagnostic performance of fluorescence in situ hybridization (FISH), nuclear matrix protein 22 (NMP22), and their combination model in bladder carcinoma detection: a systematic review and meta-analysis. Onco. Targets. Ther. **12**, 349–358 (2019)
128. Hajdinjak, T.: UroVysion FISH test for detecting urothelial cancers: meta-analysis of diagnostic accuracy and comparison with urinary cytology testing. Urol. Oncol. **26**, 646–651 (2008)
129. Horstmann, M., Patschan, O., Hennenlotter, J., Senger, E., Feil, G., Stenzl, A.: Combinations of urine-based tumour markers in bladder cancer surveillance. Scand. J. Urol. Nephrol. **43**, 461–466 (2009)
130. Todenhofer, T., Hennenlotter, J., Esser, M., Mohrhardt, S., Tews, V., Aufderklamm, S., Gakis, G., Kuehs, U., Stenzl, A., Schwentner, C.: Combined application of cytology and molecular urine markers to improve the detection of urothelial carcinoma. Cancer Cytopathol. **121**, 252–260 (2013)
131. He, H., Han, C., Hao, L., Zang, G.: ImmunoCyt test compared to cytology in the diagnosis of bladder cancer: a meta-analysis. Oncol. Lett. **12**, 83–88 (2016)
132. Shao, C.H., Chen, C.L., Lin, J.Y., Chen, C.J., Fu, S.H., Chen, Y.T., Chang, Y.S., Yu, J.S., Tsui, K.H., Juo, C.G., et al.: Metabolite marker discovery for the detection of bladder cancer by comparative metabolomics. Oncotarget. **8**, 38802–38810 (2017)
133. Liu, X., Cheng, X., Liu, X., He, L., Zhang, W., Wang, Y., Sun, W., Ji, Z.: Investigation of the urinary metabolic variations and the application in bladder cancer biomarker discovery. Int. J. Cancer. **143**, 408–418 (2018)
134. Loras, A., Trassierra, M., Sanjuan-Herraez, D., Martinez-Bisbal, M.C., Castell, J.V., Quintas, G., Ruiz-Cerda, J.L.: Bladder cancer recurrence surveillance by urine metabolomics analysis. Sci. Rep. **8**, 9172 (2018)
135. Wittmann, B.M., Stirdivant, S.M., Mitchell, M.W., Wulff, J.E., McDunn, J.E., Li, Z., Dennis-Barrie, A., Neri, B.P., Milburn, M.V., Lotan, Y., et al.: Bladder cancer biomarker discovery using global metabolomic profiling of urine. PLoS One. **9**, e115870 (2014)

6 Urinary Markers of Podocyte Dysfunction in Chronic Glomerulonephritis

Natalia Chebotareva, Irina Bobkova, Lidia Lysenko, and Sergey Moiseev

Abstract

Chronic glomerulonephritis (CGN) is a disease with a steady progressive course that involves the development of nephrosclerosis, which is especially evident in clinical courses with incidences of high proteinuria (PU). Currently, proteinuria is considered the main laboratory feature (sign) of CGN activity and progression because proteinuria is closely related to the process of tubulointerstitial fibrosis, which is correlated with the grade of renal insufficiency. The injury to podocytes, which are key components of the filtration barrier, plays a central role in proteinuria development. The detachment of podocytes from the glomerular basement membrane leading to podocytopenia is suggested to induce glomerulosclerosis and hyalinosis with obliteration of capillary loops and the progression of chronic kidney disease. Urinary markers of podocyte dysfunction could serve as useful tools while monitoring the activity and prognosis of CGN. In this chapter, the most important mechanisms of podocyte loss and urinary markers of this process are discussed.

Keywords

Chronic glomerulonephritis · Urinary biomarkers · Podocyturia · Nephrinuria · Heat shock proteins · VEGF · Matrix metalloproteinases · WT-1 · Podocytopenia

6.1 Introduction

The modern concept of the pathogenesis of chronic glomerulonephritis involves an interaction between damaging and anti-damaging/anti-inflammatory factors. Damage to podocytes, which are key components of the filtration barrier, plays a central role in glomerular lesions and associated disorders of self-protection leading to the development of proteinuria, which is still considered the clinical equivalent of chronic glomerulonephritis (CGN) activity/progression [1–5].

6.2 Podocyte Damage. Podocyte Damage Markers in Chronic Glomerulonephritis

In chronic glomerulonephritis, inflammation initially affects glomeruli and progresses because of the limited regenerative capacity of glomeruli. This feature is primarily related to podocytes,

N. Chebotareva (✉) · I. Bobkova
L. Lysenko · S. Moiseev
Tareev Clinic, Department of Nephrology, Sechenov First Moscow State Medical University, Moscow, Russia

H. M. Baptista Carreira dos Santos (ed.), *Translational Urinomics*, Advances in Experimental Medicine and Biology 1306, https://doi.org/10.1007/978-3-030-63908-2_6

which are the epithelial cells of the glomerular filter [6, 7]. At birth, few podocytes are laid at approximately 800 cells per glomerulus in 2 million nephrons. In adults, some podocytes are constantly lost, but the number of podocytes suffices to maintain the filter barrier function. Kidney diseases, particularly glomerulonephritis, accelerate podocyte damage and depletion. If the number of podocytes in a glomerulus decreases by over 60% of the norm, the glomerulus becomes sclerosed [8].

6.2.1 Glomerular Filter Structure and Function

The glomerular filtration barrier comprises a capillary basement membrane, a monolayer of endothelial cells lining the glomerular filter from the inside, and a monolayer of podocytes that cover the glomerular filter from the urinary space [9]. Podocytes and the podocyte slit diaphragm play an important role in the permeability mechanisms of the glomerular filter and form the main barrier for protein passage [10]. In addition to glomerular permeability, podocytes perform several important functions as follows: podocytes participate in the maintenance of the capillary loop structure by restructuring the actin cytoskeleton, counteract the intra-capillary hydrostatic pressure, provide endocytosis of filtered proteins and immunoglobulins that can damage the glomerular filter, participate in collagen type IV synthesis, basement membrane repair, and the immune response, and support the normal function of other glomerular cells (mesangial and endothelial cells) through the production of vascular endothelial growth factor (VEGF) [11, 12].

6.2.2 Podocyte Response to Damage. Podocyturia

Podocyte damage disrupts the glomerular basement membrane (GBM) according to the location of these cells outside the capillary wall, leading to proteinuria and nephrotic syndrome [13]. The earliest sign of podocyte damage, which is only detected by electron microscopy, is the podocyte foot process effacement. This damage spreads by the reorganization of the actin cytoskeleton and redistribution of actin microfilaments, resulting in the loss of the shape of the podocyte processes. After mechanical stretch, the podocyte foot processes efface, and the podocytes detach from the GBM and desquamate [14]. Due to podocyte foot process effacement, the slit diaphragm is stretched, the intercellular space increases, serum proteins enter the urinary space, and proteinuria develops [1–5].

During the process of the spread, the cells lose their usual form, tightly and directly attach to the GBM, and fill the subpodocytic space.

In most cases, podocyte foot process effacement precedes the cell detachment from the basement membrane. GBM denudation under the high intraglomerular pressure and adhesion of glomerular loops to the Bowman's capsule forms "hernia-like" protrusions [15]. These processes both increase protein loss and trigger glomerular sclerosis.

The role of podocytes in the mechanisms of proteinuria development and the search for valuable podocyte dysfunction markers have been studied in recent years. Vogelmann SU et al. [16] were among the first to test the excretion of specific podocyte proteins (podocalyxin, synaptopodin, and WT-1) and find their increase in urine from CGN patients. These issues were most intensively studied by a group of researchers [17–19], who showed that the urinary excretion of podocytes and their structural proteins increases in parallel with proteinuria and reflects glomerulonephritis activity, while in healthy individuals and patients with CGN remission, only scarce podocytes were found in the urine [20–22].

Many primary glomerulopathies, including focal segmental glomerulosclerosis (FSGS), membranous nephropathy, and membranoproliferative glomerulonephritis, are accompanied by podocyte damage and podocyturia. Hara et al. also showed podocyte detachment from the GBM and desquamation in urine in an active IgA nephropathy with podocyturia reflecting the activity of the disease [23]. Podocyte damage is suggested to be secondary and result from increased renin angiotensin aldosterone system (RAAS) activity and intra-glomerular pressure. Regarding IgA nephropathy, the authors also discuss the role of complement activation and subepithelial membrane attack complex C5b-9 deposition [24]. The authors further discuss the role of the mechanical influence of changed glomerule architectonics on podocytes under primary damage to mesangial cells. In an experiment involving Thy-1 nephritis, the change in the shape of the capillary loops due to damage to mesangial cells (mesangiolysis) leads to the protrusion of the capillary loops into the urinary space. Under the resulting shear stress, the podocytes stretch and desquamate. Compared with proteinuria, podocyturia appears to be a more accurate marker of glomerular damage according to recent studies [25, 26].

Therefore, two experimental models of glomerular damage, i.e., primary (PAN-nephrosis) and secondary (anti-Thy1.1-nephritis) GN, showed podocyturia and proteinuria development at an early stage of nephritis. However, at a later stage of experimental nephritis, podocyturia (active damage) disappeared, while proteinuria persisted. The authors believe that podocyturia is a more specific marker of active glomerular damage, while proteinuria does not allow differentiation between a chronic glomerular barrier defect and active damage to the glomeruli [25, 26].

To determine the severity of glomerular filter damage in patients with various forms of glomerulonephritis, the following tests are used: cell culture to detect "viable" and apoptotic podocytes in urine; identification of various structural proteins of podocytes and the slit diaphragm, including nephrin, podocin, podoplanin, synaptopodin, and podocalyxin; anti-podocalyxin flow cytometry antibody product techniques; and urine immunoblotting to detect the mRNA of these proteins [27, 28].

6.2.3 Nephrinuria

In podocyte damage, specific podocyte-derived proteins may be found in urine, and primarily nephrin, which is the main protein of the slit diaphragm, is considered an early marker of podocyte damage. Nephrin binds actin fibres in the podocyte cytoskeleton and forms the intercellular slit diaphragm [29].

Since the slit diaphragm is a type of "signalling platform" necessary to maintain the functions of podocytes, the release of nephrin, its structural component, is also a stereotypical podocyte reaction to damage in active CGN [30–32].

Recent experimental studies have shown an increase in urinary nephrin excretion in nephrotic syndrome because of damage to podocytes and the closely related slit diaphragm [33]. In a rat PAN nephrosis model, Luimula et al. found an increase in the urine excretion of nephrin at the peak of proteinuria [34]. In experimental membranous nephropathy (Heymann nephritis), complement components (C5-9) enter the podocyte membrane, leading to F-actin damage and nephrin dissociation from the podocyte with subsequent excretion in urine [35]. Even before the development of proteinuria, electron immunomicroscopy of kidney tissue shows sites of slit diaphragm destruction corresponding to areas exhibiting reduced nephrin expression. In advanced disease, if high proteinuria develops, the number of these areas significantly increases, with an uneven distribution and alternation with the completely preserved areas of the slit diaphragm [36]. These studies showed for the first time that the relationship between nephrinuria and focal slit diaphragm defect development is an important mechanism of proteinuria (PU) [37]. The data reveal a decrease in nephrin expression in kidney tissue regardless of the morphological form (minimal change disease, FSGS, membranous nephropathy, MPGN, or IgAN) [38–40].

We tested the podocytic diaphragm lesion in patients with chronic glomerulonephritis by nephrin excretion in urine (nephrinuria). In total, 73 CGN patients were studied as follows: 20 patients with inactive CGN (group III); 23 patients with active CGN with proteinuria (PU) >1 g/d (group I); and 30 patients with nephrotic syndrome (NS) (group II), including 11 patients with severe NS (PU >10 g/d, hypoalbuminemia <20 g/L) (IIb) and 19 patients with moderate NS (IIa) (Table 6.1).

Table 6.1 Subjects characteristics

Groups value	Group I n = 23	Group II (NS) n = 30	Group IIa moderate NS n = 19	Group IIb severe NS n = 11	Group III n = 20
Proteinuria, g/24 h	1.9 [1.4–3.2]	6.12 [4.37–10.0]	5.54 [3.4–7.9]	10.0 [7.3–16.0]	0.48 [0.157–0.78]
Serum albumin, g/l	39.8 [37.5–42.4]	25.6 [21.3–30.7]	28.7 [25.0–31.0]	19.0 [17.5–20.0]	43.8 [41.2–46.1]
Serum total protein, g/l	63.2 [59.7–68.7]	43.9 [39.2–50.6]	46.9 [42.5–51.6]	37.9 [33.6–37.7]	67.8 [65.2–71.9]
Serum creatinin, mg/dl	1.1 [0.85–1.65]	1.06 [0.8–1.5]	1.0 [0.72–1.4]	1.25 [0.9–1.93]	1.03 [0.84–1.5]
eGFR CKD-EPI ml/min/1.73 m^2	87.3 [55.0–114.5]	90.0 [60.0–117.0]	101.0 [76.0–134.0]	65.3 [43.0–98.8]	76.0 [55.0–116.0]
Systolic blood pressure, mm Hg	130 [125–140]	135 [120–160]	140 [120–160]	135 [130–150]	120 [120–135]
Diastolic blood pressure, mm Hg	80 [80–90]	90 [80–100]	90 [80–100]	83 [80–90]	80 [80–90]

Abbreviations: *NS* nephrotic syndrome, *eGFR* estimated glomerular filtration rate, CKD-EPI formula

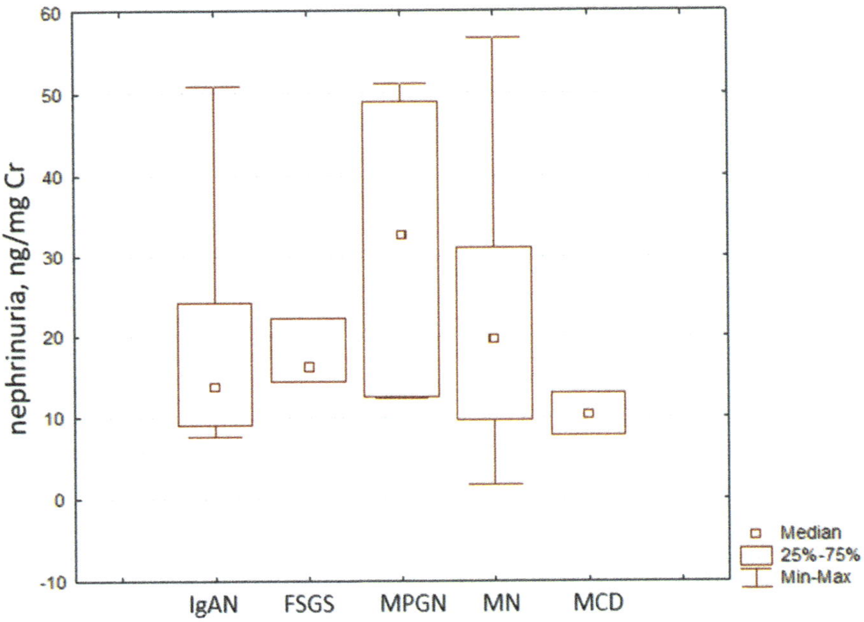

Fig. 6.1 Nephrinuria in patients with various forms of CGN. Abbreviations: *IgAN* Immunoglobulin A nephropathy, *FSGS* focal segmental glomerular sclerosis, *MN* membranous nephropathy, *MPGN* membranoproliferative glomerulonephritis, *MCD* minimal change disease

Additionally, 8 healthy subjects were studied as a control group. The levels of nephrin in the urine were assessed by ELISA. The levels of nephrinuria in the patients with active CGN were significantly higher than the levels in the patients who were in remission and the levels in the healthy individuals. These levels in the patients with NS were significantly higher than those in the patients with lower levels of proteinuria. The highest urinary levels of nephrin were observed in the patients with heavy nephrotic syndrome and renal dysfunction.

The differences between the morphological forms of both primary "true" podocytopathies (minimal change disease, FSGS, and membranous nephropathy) and other proliferative forms of CGN (MPGN and IgAN) were insignificant in terms of nephrinuria; the severity of proteinuria and nephrotic syndrome played a decisive role (Fig. 6.1).

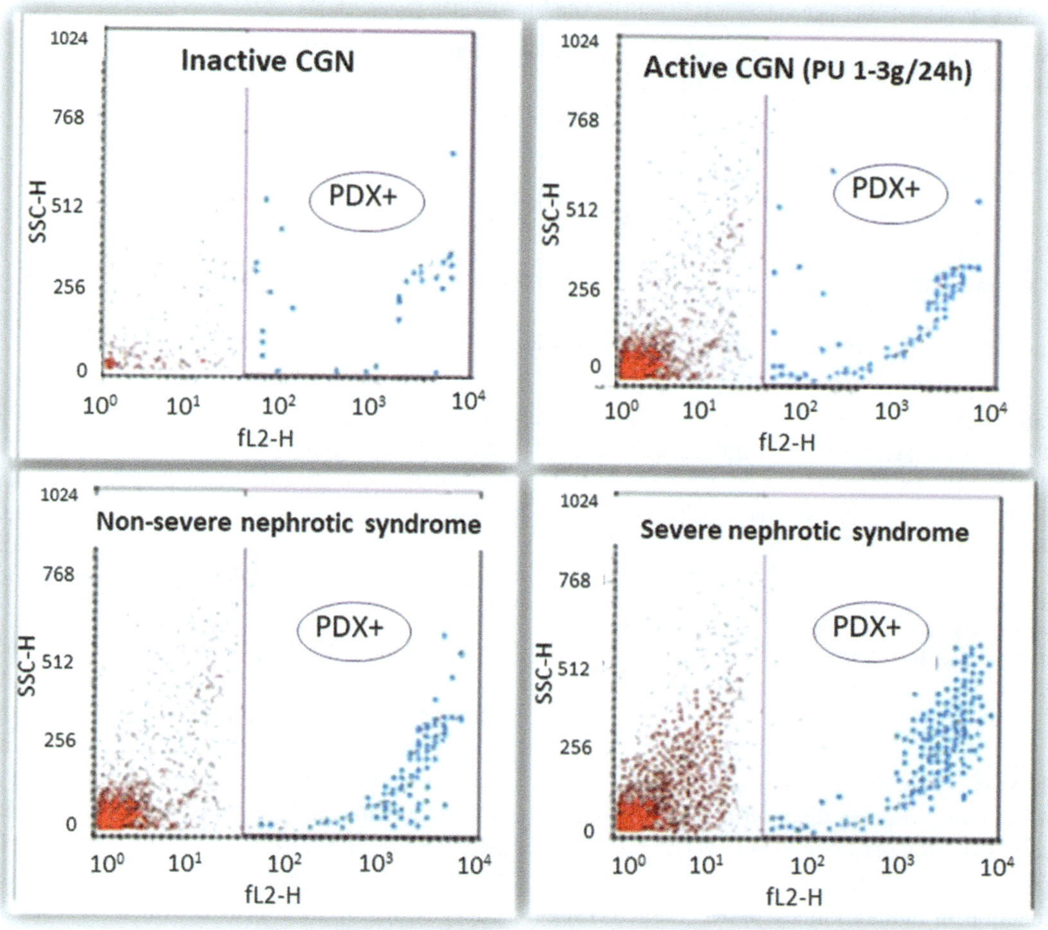

Fig. 6.2 Urine cytoflowgrams in patients with different levels of CGN activity

Active CGN manifested not only by the urinary excretion nephrin, which is a protein that plays structural and functional roles, but also podocytes because of their detachment from the glomerular basement membrane and desquamation of damaged podocytes in urine (Fig. 6.2). Podocyturia was assessed by flow cytometry, which applied phycoerythrin-conjugated, mouse monoclonal, and anti-human antibodies to the podocyte surface antigen podocalyxin. The excretion of whole cells or cell fragments in 100 mL increments of urine sediment was detected. Among the patients with nephrotic syndrome, podocyturia was significantly higher than that in the other groups (Table 6.2). Podocyturia was correlated with proteinuria ($Rs = 0.23$, $p < 0.05$).

Regarding nephrinuria, we found significant correlations between podocyturia and proteinuria, serum albumin, and duration of nephrotic syndrome, highlighting the increasing severity of podocytic damage in patients with NS (Tables 6.2 and 6.3). Podocyturia was closely correlated with nephrinuria ($Rs = 0.9$, $p < 0.05$).

6.3 Heat Shock Protein 27

In addition to the "classical" markers, we studied a small heat shock protein 27 as an alternative marker of podocyte damage and disturbance to the glomerular filter.

Small (low-molecular) HSP is a group of proteins that vary in molecular weight from 16 to

Table 6.2 Nephrinuria and podocyturia in patients with chronic glomerulonephritis

Study group	n	Podocyturia (pdx + cells/μL)	Nephrinuria (ng/mg Cr)
Healthy subjects	10	0[0;0.6]	7.9 [1.7; 9.5]
CGN group I (prominent urinary syndrome)	23	5.45 [2.5;18.3]	9.5 [7.6; 13]*
CGN group II	38	12.0 [7.0;35.0] **	16.9 [10.9; 27.8] **
(ns) Moderate NS	28	8.0 [4.2;15.5]	16.6 [10.45;27.3]
Severe NS	10	36.5 [9.0; 80.3]	17.8 [16.1;37.5]
CGN group III (inactive)	20	5.4 [1.0;9.43]	7.25 [5.4; 9.8]

*p<0.05 group II vs. group I and healthy; p<0.05 significant NS vs moderate NS, **p < 0.01 vs. Groups I and II

Table 6.3 Relation of podocyte damage parameters (podocyturia, nephrinuria) with clinical and laboratory signs of CGN activity

Parameter	Podocyturia	Nephrinuria
Proteinuria, g/d	Rs = 0.23. p < 0.05	Rs =0.27. p < 0.05
Blood serum albumin, g/L	Rs = −0.38. p < 0.001	Rs = −0.27. p < 0.05
NS duration, mo	Rs = 0.42. P < 0.05	Rs = 0.4. p < 0.05

40 kDa. These proteins play an important role in maintaining normal function and stabilizing the actin cytoskeleton of cells [41, 42]. Among these proteins, HSP27 performs an important protective function in human kidney tissue. Larger HSP27 molecules coexist in the cell as chaperone proteins, and smaller HSP27 oligomers associated with actin fibres (actin-associated proteins) stabilize actin filaments if damaged by proinflammatory factors [43–45]. These proteins confer the main protective role to unphosphorylated HSP27 oligomers bound to actin microfilaments after phosphorylation by p38MAP kinase and remodelling of the cell actin network.

The increase in unphosphorylated HSP27 blocks the restructuring of the actin cytoskeleton, stabilizes the cytoskeleton, and increases the resistance of the cell to damage. HSP27 in kidney glomeruli is localized almost entirely in podocytes with a well-developed actin cytoskeleton. The mechanisms of podocyte actin cytoskeleton damage under the influence of proinflammatory factors are under investigation. Podocyte damage by various pro-inflammatory factors is supposedly accompanied by adenosine triphosphate (ATP) deficiency in the cell, unphosphorylation of adhesive proteins, and aggregation (or destruction) of actin microfilaments in the cytoskeleton [46]. Another cause of podocyte damage under ATP deficiency is the primary inherited pathology of its actin cytoskeleton [47].

Under physiological conditions, HSP27 is not associated with actin fibres and is located intracellularly. When the cell is damaged, unphosphorylated HSP27 forms closer bonds with cellular actin, thus stabilizing the intracellular microfilaments and cytoskeleton of podocytes and tubular cells in the kidney, rendering the cells more resistant to stress [45, 48]. HSP27 phosphorylation by mitogen activated kinases is accompanied by the separation of HSP27 from actin microfilaments, remodelling of the actin network, aggregation and redistribution of microfibres in the cell, and podocyte foot process effacement; therefore, HSP27 plays an important role in maintaining the integrity of podocyte processes during damage. Any alterations in actin microfilaments due to podocyte stress may lead to the accumulation of HSP27, the polymerization of actin and changes in the podocyte shape, which is a process known as podocyte foot process effacement.

Smoyer et al. showed a significant increase in BTS27 expression in glomeruli in rats. The induction of nephrotic syndrome led to an increase in glomerular HSP27 expression and phosphorylation [49, 50].

An increase in the glomerular expression of HSP27 was observed in nephropathies manifested by proteinuria and nephrotic syndrome, i.e., diabetic nephropathy and FSGS, and in patients with hypertensive nephropathy. Tsagalis et al. showed the important role of HSP27 in intracellular protection against lupus nephritis in humans. Increased HSP27 expression is observed mainly in resident kidney cells (glomerular and tubular). The HSP27 expression level in kidney tissue was high in diffuse proliferative nephritis and correlated with the disease activity parameters and serum creatinine levels [51].

In our study, stress to podocytes in patients with CGN with high proteinuria (Groups I and II) resulted in an increase in the urine HSP27 levels. HSP27 excretion was the highest in the patients with severe NS (Group II), less significant in the patients with subnephrotic proteinuria (Group I), and insignificant in the patients with CGN remission (Group III) (Table 6.4). The urine level of protective HSP27 is directly correlated with the severity of PU ($Rs = 0.297$, $p < 0.05$), which is explained by the data showing that the role of this protein in podocyte foot process changes leads to PU.

Table 6.4 HSP27 urinary excretion in CGN patients (n = 84)

Parameter Group	HSP27 (ng/mg Cr)
Healthy (n = 10)	0.73 [0.66; 0.96]
I. CGN with urinary syndrome (n = 34)	0.76 [0.68; 1.14]
II. CGN with NS (n = 30)	1.1 * [0.73; 1.83]
Moderate NS (n = 23)	0.94 [0.71; 1.73]
Significant NS (n = 7)	1.88 *[0.8; 7.195]
III. CGN remission (n = 10)	0.72 [0.65; 0.98]

*p<0.05 group II vs. group I and healthy; p<0.05 significant NS vs moderate NS

Since the content of heat shock protein 27 in a cell under the influence of damaging factors explodes, the severe damage to podocytes in patients with NS is accompanied by a significant depletion of these cells and the release of intracellular proteins in the urine. Thus, an increase in urine HSP27 can indicate both increased damage and the violation of podocyte self-protection.

In patients with NS, the three parameters studied (podocyturia, nephrinuria, and urine HSP27 level) are correlated, confirming both their value as pathogenically related markers of podocyte dysfunction, which is an important component in the development of proteinuria, and a disturbance of the kidney damage/self-protection (heat shock proteins) systems interaction in podocyte territory (Fig. 6.3).

6.4 Factors of Apoptosis and Survival of Podocytes in Urine from CGN Patients

An important component of the mechanism of podocyte depletion and increase in proteinuria is an increase in podocyte apoptosis, which is marked by the activation of the intracellular caspase cascade. Urine caspase-9 excretion has been shown to be significantly higher in patients with active forms of CGN with NS than in those with urinary syndrome. The caspase-9 level increases in parallel with the increase in nephrinuria, showing the contribution of apoptosis to podocyte damage, which leads to ne phrinuria and proteinuria development in CGN patients (Fig. 6.4).

Fig. 6.3 Relation of podocyte lesion parameters in CGN patients with NS

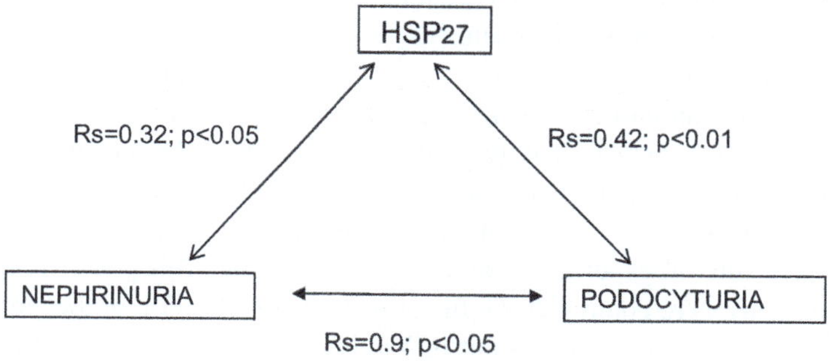

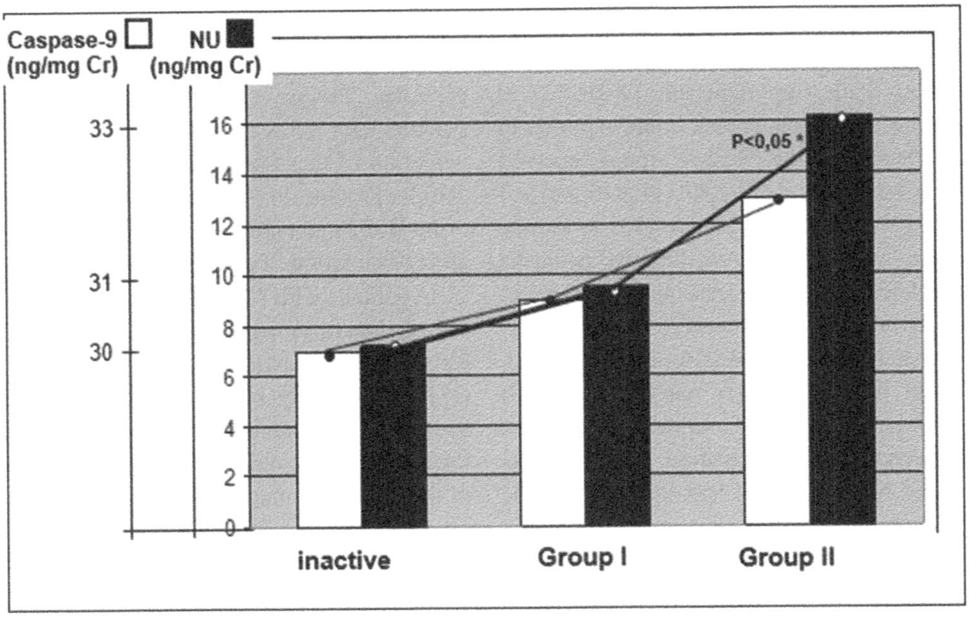

Fig. 6.4 Caspase-9 and nephrin in the urine of CGN patients

6.4.1 Vascular Endothelial Growth Factor (VEGF) in CGN

The urinary excretion of antiapoptotic Vascular Endothelial Growth Factor (VEGF) further indicates a decrease in podocyte survival.

VEGF is an angiogenic factor and a key regulator of the glomerular filtration barrier [52]. An experiment showed the development of proteinuria in mice treated with anti-VEGF antibodies or soluble VEGF receptor 1 (sFlt-1) [53]. Oncological patients undergoing anti-VEGF therapy experience proteinuria and podocyturia [54–57]. Women with preeclampsia, which is characterized by the circulation of sFlt-1 of placental origin, develop anti-VEGF treatment-like symptoms [58, 59]. These data confirm the important role of VEGF in regulating the glomerular barrier and the mechanisms of proteinuria development.

In human kidneys, the most common form is VEGF-A, which is mainly expressed by podocytes and partially expressed by tubular cells [60]. In different forms of glomerulonephritis, the results of experiments and a study involving humans were contradictory. In some experimental CGN models, the amount of VEGF was elevated and correlated with the proteinuria levels [61]. In minimal change CGN, the urinary VEGF level was also elevated and depended on the severity of NS [62], and a study investigating VEGF expression in kidney tissue showed its increase primarily in podocytes [63]. Hohenstein et al. found an increase in VEGF expression in kidney biopsy from patients with endocapillary nephritis, MCGN, and crescentic glomerulonephritis. In kidney biopsies from patients with membranous nephropathy, VEGF was also expressed mainly in podocytes. In contrast, other experimental and clinical studies have shown a decrease or absence of VEGF expression in FSGS, diabetic nephropathy, crescentic glomerulonephritis, and lupus nephritis [64–68].

The results indicate the complex regulation of the VEGF signal necessary to maintain the glomerular filtration barrier, including the normal phosphorylation of nephrin and the slit diaphragm protein. Violation of nephrin phosphorylation in VEGF deficiency weakens nephrin binding to podocin, leading to a separation of the extracellular part of the nephrin molecule from the podocyte membrane, and increases the excretion of this protein in urine. Damage to the endothelium and podocytes with PU development in

VEGF deficiency were first shown in women with pre-eclampsia [69]. In anti-GBM nephritis, VEGF deficiency formation exacerbates the disease course severity with impairment of the slit diaphragm protein function, loss of podocyte foot processes, and proteinuria development [70]. We obtained evidence indicative of a decrease in the survival of podocytes based on a study investigating the urinary excretion of the anti-apoptotic factor VEGF. VEGF is produced by podocytes and supports the function of nephrin and the slit diaphragm while also preserving the integrity of the glomerular barrier.

In our study, the urinary levels of VEGF were the highest in the patients with NS but were observed to decline when renal dysfunction appeared. However, the nephrinuria levels continued to rise as a result of the severity of podocyte dysfunction. A decline in VEGF levels under the conditions of NS and renal dysfunction may indicate the depletion of podocytes (Fig. 6.5).

Urinary VEGF in patients with active CGN with preserved renal function (Groups I and II) is directly correlated with the PU ($Rs = 0.673$ $p < 0.0001$), and in patients with progressive disease and renal dysfunction, the urinary VEGF reduction depended on the serum creatinine level ($Rs = -0.31$ $p < 0.05$). We found a negative correlation between urine VEGF excretion and the severity of nephrinuria ($Rs = -0{,}8$, $p < 0.05$).

6.5 Mechanisms of Podocytopenia Are the Main Determinants of Glomerulosclerosis

Podocyte desquamation and depletion are the central mechanisms of glomerular sclerosis in glomerulonephritis [71]. There are several hypotheses related to podocytopenia development in immune-mediated diseases.

Podocyte damage leads to a disruption of adhesive protein synthesis in podocytes, the separation of these cells from the glomerular basement membrane and, ultimately, their desquamation into the urinary space, i.e., so-called podocyturia. Podocytes, which are highly differentiated cells, are incapable of proliferation and substitution; therefore, the progressive depletion of these cells in the glomerule leads to GBM denudation and triggers glomerulosclerosis [72, 73].

Apoptosis is another equally important mechanism of podocyte depletion from GBM and excretion into the urinary space [74, 75]. This role is discussed in podocyte hypertrophy and the epithelial-mesenchymal transition, which reduce the adhesive properties of cells and promote the desquamation of "living" podocytes [76–78]. Studies involving cell culture have shown that the response of podocytes to mechanical stretch is hypertrophy, particularly by activating cell cycle

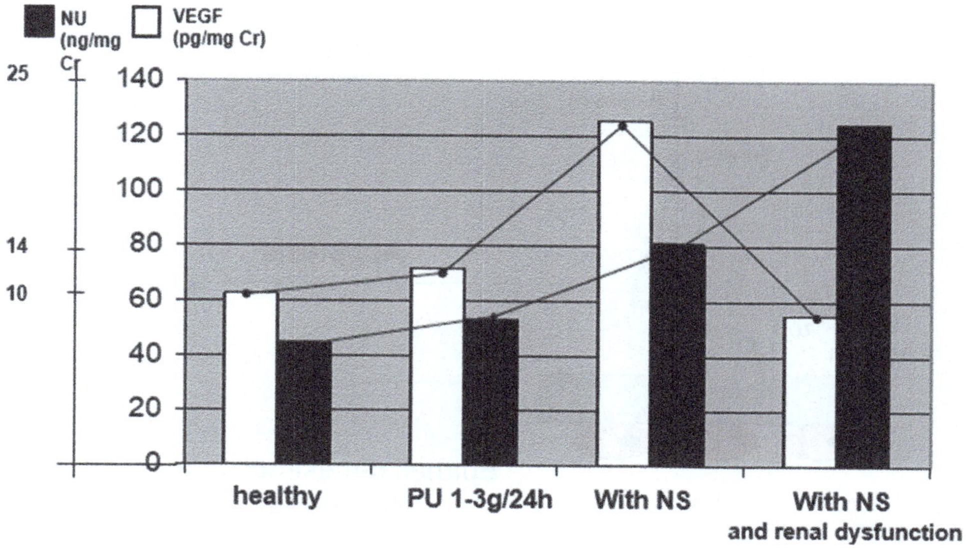

Fig. 6.5 VEGF and nephrin urinary excretion in CGN patients

inhibitors [79], which is often accompanied by nucleus division without the complete division of the cell as illustrated by the appearance of dual- and multi-nucleated podocytes [80]. Podocytes adhere less to GBM and desquamate in urine [81] (Fig. 6.6).

Under these conditions, the intensive podocyte detachment from the GBM and their depletion in urine lead to podocytopenia accompanied by the "denudation" of separate areas of the GBM, loss of the capillary loop shape, and formation of local basement membrane protrusions. The contact of these protrusions with glomerular parietal cells forms synechiae between the capillary loops and Bowman's capsule, starting glomerular sclerosis [73]. Since podocytes are highly differentiated cells with limited proliferative potential, a decrease in the number of podocytes leads to an irreversible defect in the glomerular filter, contributes to the formation of synechiae between the capillary loops and Bowman's capsule, and starts glomerular sclerosis (Fig. 6.7). A critical decrease in the number of podocytes (podocytopenia) in a glomerule is the main factor in the progression of glomerular sclerosis and decrease in the glomerular filtration rate [74, 75]. In 2005, Wharram BL et al. used a glomerulosclerosis model in rats to show that an approximately 20% depletion of podocytes per glomerule leads to the development of transient proteinuria and that a total depletion >40% of podocytes is accompanied by persistent high proteinuria with impaired kidney function; morphologically, multiple synechiae formation between the capillary loops and Bowman's capsule, focal segmental sclerosis development, and global glomerulosclerosis are detected during the final stage [82].

6.6 Immunohistochemical Evaluation of Podocytopenia in Patients with Chronic Glomerulonephritis

We also confirmed the progressive loss of podocytes in glomeruli in the active types of CGN with proteinuria by the use of immunohistochemistry.

Compared with the control group, which was represented by individuals with normal renal tissue removed via nephrectomies in cases of renal carcinomas, a significant decrease in the expression level of WT-1 in the glomeruli in this group of patients was established.

The number of WT-1 nuclei expressing WT-1 in the patients with active CGN with NS was reduced (21 [16, 27.5] cells/section of the glomerulus) compared with that in the control group (44 [38; 47] cells/section of the glomerulus) (Fig. 6.8).

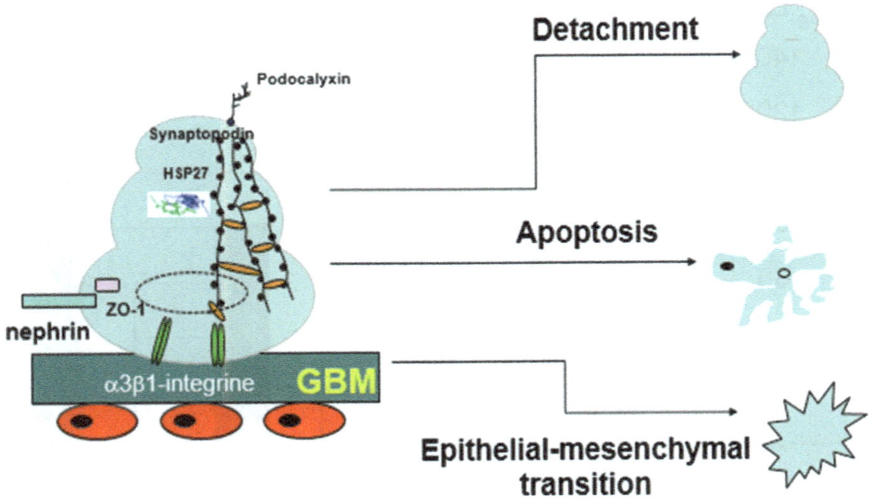

Fig. 6.6 Mechanisms of podocyte loss

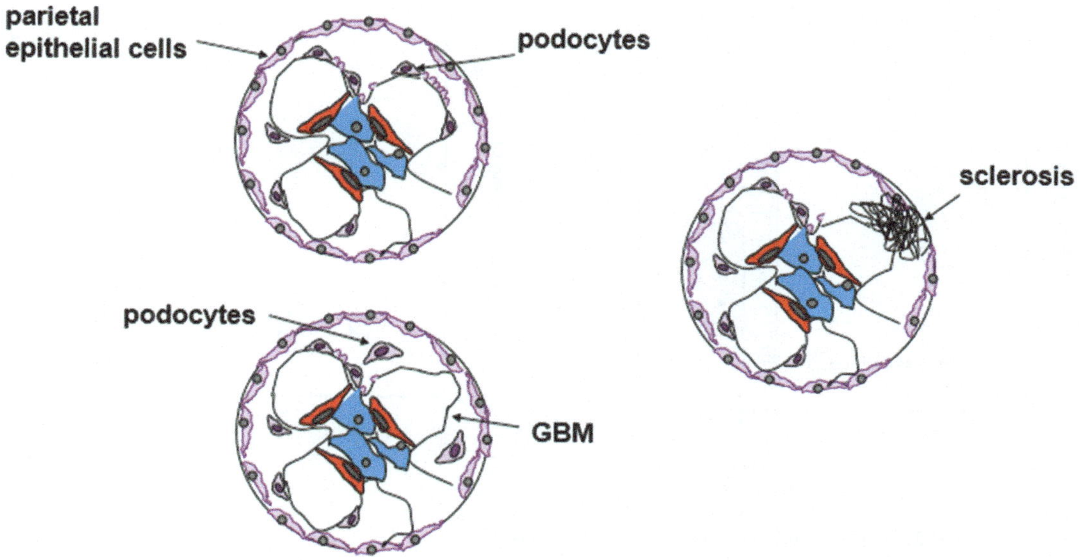

Fig. 6.7 Mechanism of glomerulosclerosis - podocytopenia

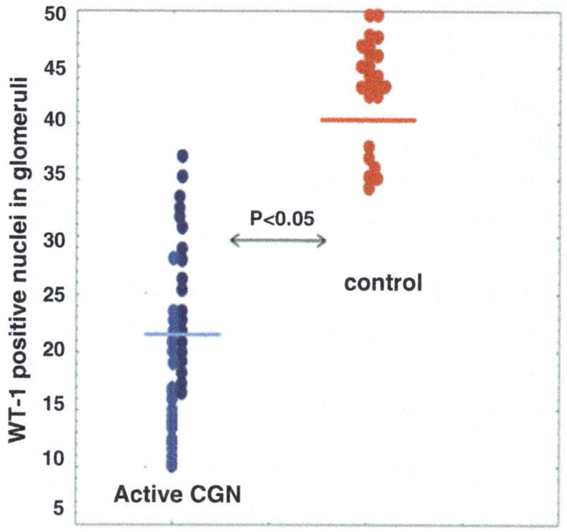

Fig. 6.8 WT-1 expression to test the podocytopenia in CGN patients

A correlation was found between podocyturia and proteinuria (Rs = −0.35, $p < 0.05$) and the severity of podocytopenia (Rs = −0.47, $p < 0.05$), which is a determinant of glomerulosclerosis.

6.7 Matrix Metalloproteinases and Their Inhibitor Disturbance

Matrix metalloproteinases (MMP) constitute a large family of Zn-containing and Zn-dependent endopeptidases; approximately 28 different MMP are known. MMP function mainly to control extracellular matrix component cleavage and accumulation, but in severe inflammation, to a great extent, MMP perform a pro-inflammatory function by activating cytokines, growth factors, and adhesive molecules [83–86]. MMP contribute to the epithelial-mesenchymal transition (EMT) of kidney tissue cells (podocytes and tubular cells). In the course of the EMT, podocytes are de-differentiated and acquire mesenchymal cell markers [87]. MMP-2 and MMP-9 are localized in the EMT zones of the tubular epithelium and induce myofibroblasts [88, 89]. Increased MMP-2 expression in the proximal tubular epithelium stimulates GBM destruction, tubular atrophy, and myofibroblast-transformed cell migration into the kidney interstitial tissue [90].

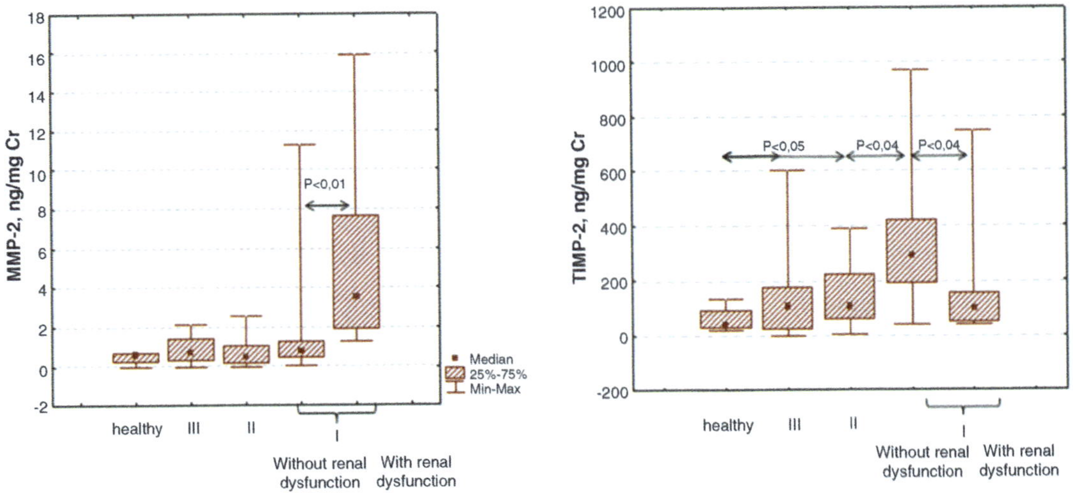

Fig. 6.9 MMP-2 and TMP-2 urine excretion in CGN patients

MMR-2 is produced by fibroblasts, epithelial cells, and mesangial cells. In glomerulonephritis, an MMP increase is associated with disease activity and inflammatory cell attraction; the level and duration of the MMR increase determine the glomerular damage burden. Despite the proteolytic properties, MMP are more likely to exert a pro-inflammatory and profibrogenic effect and induce the synthesis of profibrogenic cytokines, particularly TGF-beta [79, 91].

Metalloproteinase activity is controlled by endogenous tissue inhibitors (TIMP). TIMP (21–28 kDa) are specific endogenous inhibitors linking MMP. Among these inhibitors, four members of the family (TIMP-1,-2,-3, and -4) have been identified in humans. MMP-2 and MMP-9 form proenzyme complexes with the endogenous inhibitors TIMP-2 and TIMP-1, respectively. An increase in the TIMP/MMP ratio indicates more complete control of the proteolytic activity of metalloproteinases [92].

MMP and TIMP increases have also been found in various forms of glomerulonephritis in humans, including rapidly progressing (an ANCA-associated vasculitis), IgA nephropathy, and acute post-infection glomerulonephritis [93–96]. In the membranous nephropathy model, MMP-9 expression increased in the glomeruli, mainly in podocytes [97].

In patients with NS and renal dysfunction, the kidney self-protection/damage imbalance leading to the activation of inflammation and the epithelial-mesenchymal transition was evidenced by a violation of the urine ratio of proinflammatory MMP-2 and tissue MMP-2 inhibitor (TIMP-2). Patients with renal dysfunction showed an increase in the urinary excretion of MMP-2 with a simultaneous decrease in that of its inhibitor TIMP-2 (Fig. 6.9).

The balance between the synthesis and degradation of extracellular matrix (ECM) components is the basis of the structural and functional integrity of the glomerule. Changes in MMP expression or activity are directly involved in the intensity of ECM metabolism and may contribute to glomerulosclerosis and decrease in renal function [98].

The epithelial-mesenchymal transition (EMT) is the third way to achieve podocyte loss. In addition to inflammation, matrix metalloproteinases (MMPs) stimulate the EMT of podocytes with increased mobility, loss of contact of cells with GBM and finally detachment of cells from the GBM. We examined the expression of MMP-9 in renal tissues from patients with CGN. An increase in the MMP-9 expression levels in the glomeruli was observed in the active types of CGN. High expression levels were associated with high

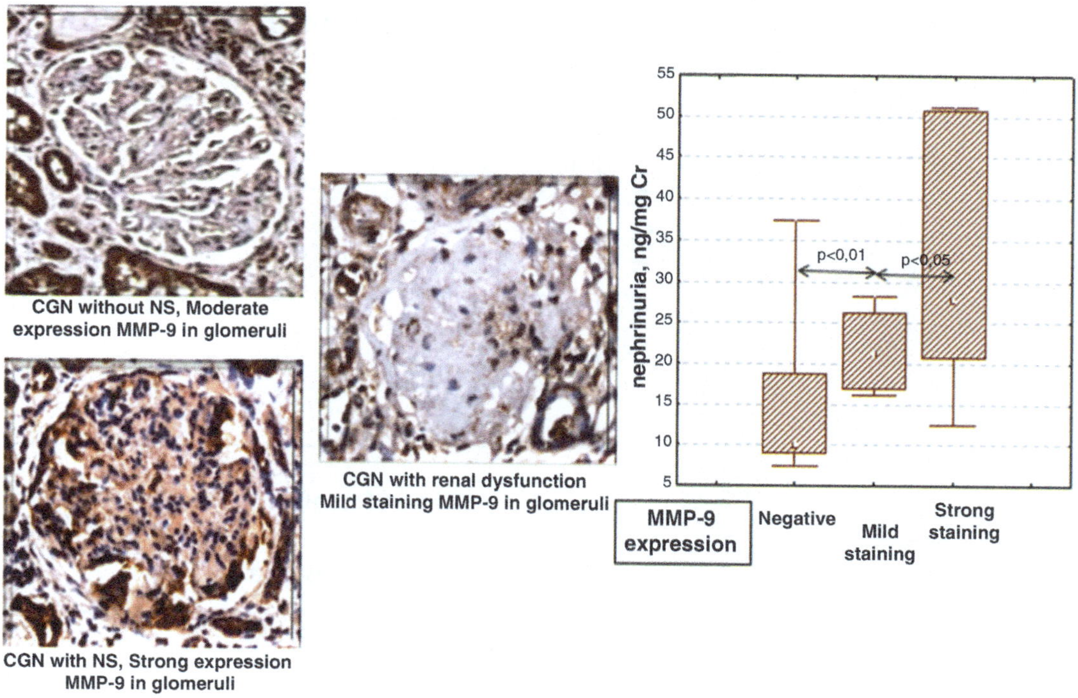

Fig. 6.10 Matrix metalloproteinase-9 expression in renal tissue and nephrinuria in CGN patients

nephrinuria levels, thus suggesting the important role of the EMT in the loss of podocytes and the development of podocytopenia (Fig. 6.10).

6.8 Prognostic Value of the Studied Parameters in CGN Patients

We also analysed the effects of immunosuppressive therapy according to the podocyturia, nephrinuria and urinary VEGF levels at baseline in 52 patients. The response to the therapy was significantly worse in the patients with high nephrinuria and podocyturia levels and those with low VEGF levels. In most patients with high baseline levels of podocyturia and nephrinuria, NS was resistant to immunosuppressive therapy, despite the long-term use of different immunosuppressive schemes of treatment, which ranged from 9 months to 3 years. The relative risk of resistance to therapy was 5 times higher under the podocyturia condition. Thus, the initial level of these parameters allows for the prediction of the response to therapy in CGN patients (Fig. 6.11).

Moreover, the accuracies of the areas under the curve of the nephrinuria and podocyturia levels were higher than those of the proteinuria and serum creatinine levels in response to immunosuppressive therapy (Fig. 6.12).

6.9 Summary

In chronic glomerulonephritis with predominant proteinuria (PU) and nephrotic syndrome (NS), podocyte damage accompanies the excretion of slit diaphragm structural proteins and desquamation of podocytes in urine; the severity of these changes depends on the clinical course of CGN, primarily the PU and kidney dysfunction. The increase in proteinuria and podocyturia in patients with active CGN is accompanied by kidney podocyte apoptosis activation, which is manifested by an increase in caspase-9 in the urine. The simultaneous increase in podocyte survival

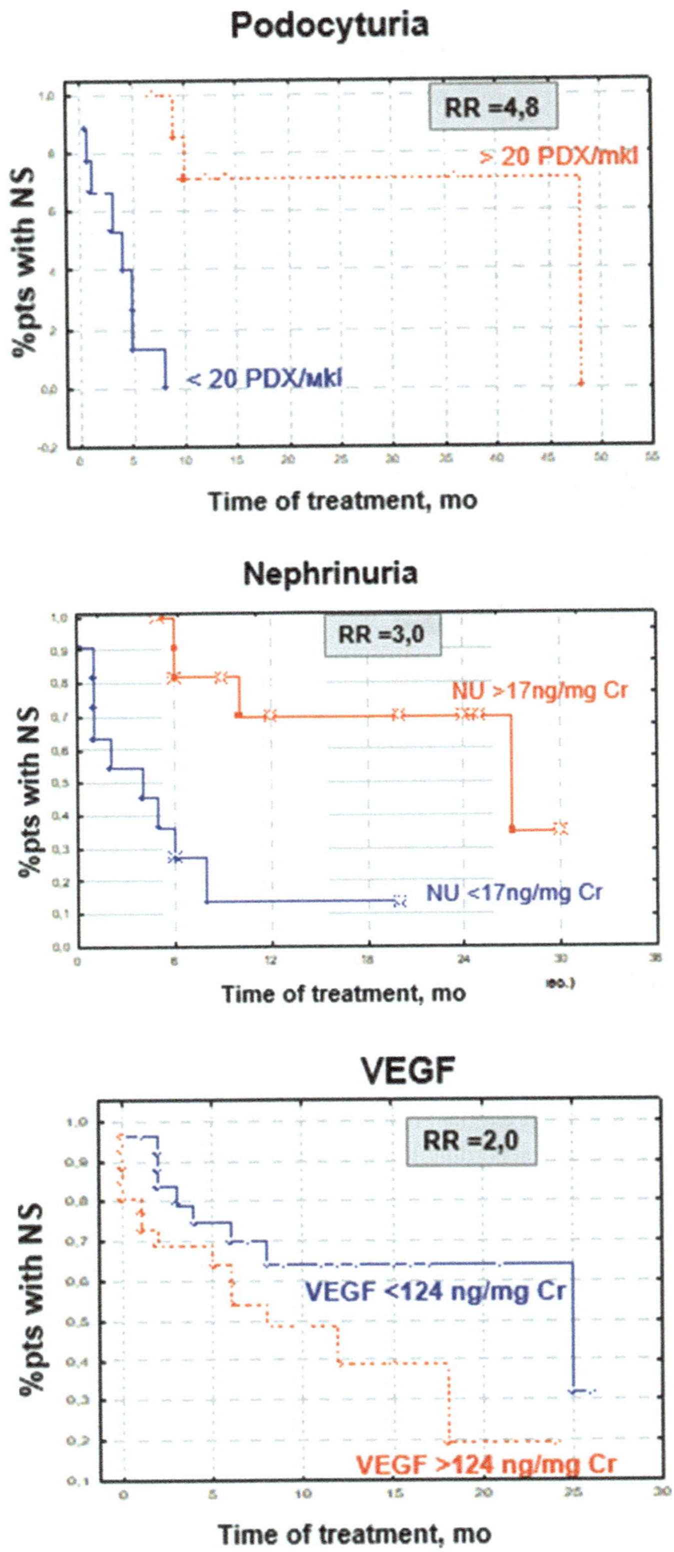

Fig. 6.11 Effect of immunosuppressive therapy in patients with active CGN (n = 55)

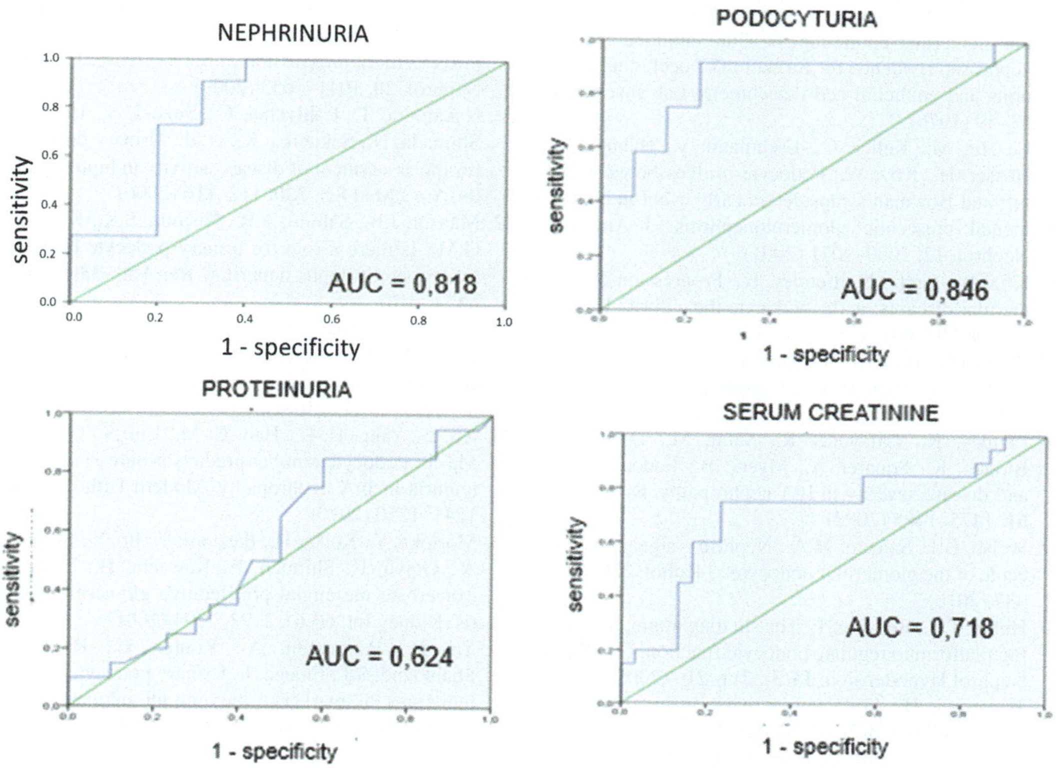

Fig. 6.12 Podocyturia, nephrinuria, proteinuria and serum creatinine in prognosis (ROC curve analysis)

factor VEGF excretion is a consequence of the compensatory activation of its synthesis; in patients with NS and renal dysfunction, VEGF production is depleted, showing the prevalence of apoptosis and progressive podocyte damage with the persisting high urine levels of caspase-9 and nephrin. In CGN with massive PU and NS, the number of podocytes in the glomeruli decreases (podocytopenia) and is correlated with podocyturia, the severity of renal dysfunction, and nephrosclerosis. In patients with NS and renal dysfunction, the podocyte damage coincides with the failure of the kidney self-protection system, which is characterized by a TIMP-2 decrease and MMP-2 increase in the urine.

The high urine levels of nephrin, podocytes, and VEGF reflect severe podocyte damage and an unfavourable prognosis. This spectrum of urinary biomarkers is accompanied by increased desquamation of damaged and dead podocytes in the urine with the bulk podocyte depletion in glomeruli (development of podocytopenia). Along with the activation of apoptosis, as a mechanism of podocyte depletion and glomerulosclerosis development, the activation of the epithelial-mesenchymal transition of damaged podocytes is of great importance. The severity of the podocyte dysfunction affects the course of CGN and the response to immunosuppressive therapy. These markers could identify patients with CGN belonging to a group at a high risk of disease progression and failure of immunosuppressive therapy.

References

1. Kriz, W., Shirato, I., Nagata, M., LeHir, M., Lemley, K.V.: The podocyte's response to stress: The enigma of foot process effacement. Am J Physiol Renal Physiol. **304**, 333–347 (2013)
2. Jefferson, J.A., Shankland, S.J., Pichler, R.H.: Proteinuria in diabetic kidney disease: A mechanistic viewpoint. Kidney Int. **74**, 22–36 (2008)
3. Patrakka, J., Tryggvason, K.: New insights into the role of podocytes in proteinuria. Nat Rev. Nephrol. **5**, 463–468 (2009)

4. Caulfield, J., Reid, J., Farquhar, M.: Alterations of the glomerular epithelium in acute aminonucleoside nephrosis. Evidence for formation of occluding junctions and epithelial cell detachment. Lab Invest. **34**, 43–59 (1976)
5. Le Hir, M., Keller, C., Eschmann, V., Hähnel, B., Hosser, H., Kriz, W.: Podocyte bridges between the tuft and Bowman's capsule: an early event in experimental crescentic glomerulonephritis. J Am Soc Nephrol. **12**, 2060–2071 (2001)
6. Kriz, W., Gretz, N., Lemley, K.: Progression of glomerular diseases: is the podocyte the culprit? Kidney Int. **54**, 687–697 (1998)
7. Pavenstadt, H., Kriz, W., Kretzler, M.: Cell biology of the glomerular podocyte. Physiol Rev. **83**, 253–307 (2003)
8. Lemley, K., Lafayette, R., Safai, M., Derby, G., Blouch, K., Squarer, A., Myers, B.: Podocytopenia and disease severity in IgA nephropathy. Kidney Int. **61**, 1475–1485 (2002)
9. Welsh, G.I., Saleem, M.A.: Nephrin—signature molecule of the glomerular podocyte? J Pathol. **220**, 328–337 (2010)
10. Huber, T.B., Bensing, T.: The slit diaphragm: a signaling platform to regulate podocyte function. Curr Opin Nephrol Hypertension. **14(3)**, 211–216 (2005)
11. Trimarchi, H.: Podocyturia: what is in a name? J Transl Int Med. **3(2)**, 51–56 (2015)
12. Barisoni, L., Schnaper, H.W., Kopp, J.B.: A Proposed Taxonomy for the Podocytopathies: A Reassessment of the Primary Nephrotic Diseases. Clin J Am Soc Nephrol. **2**, 529–542 (2007)
13. Couser, W.G.: Basic and translational concepts of immune-mediated glomerular diseases. J Am Soc Nephrol. **23**, 381–399 (2012)
14. Kriz, W., Shirato, I., Nagata, M., LeHir, M., Lemley, M.V.: The podocyte's response to stress: the enigma of foot process effacement. Am J Physiol Renal Physiol. **304**, 333–347 (2013)
15. Drumond, M.C., Deen, W.M.: Structural determinants of glomerular hydraulic permeability. Am J Physiol (Renal Fluid Electrolyte Physiol.). **266**, 1–12 (1994)
16. Vogelmann, S.U., Nelson, W.J., Myers, B.D., Lemley, K.V.: Urinary excretion of viable podocytes in health and renal disease. Am J Physiol Renal Physiol.**285(1)**, 40–48 (2003)
17. Hara, M., Yanagihara, T., Kihara, I.: Urinary podocytes in primary focal segmental glomerulosclerosis. Nephron. **89**, 342–347 (2001)
18. Hara, M., Yanagihara, T., Kihara, I., Higashi, K., Fujimoto, K., Kajita, T.: Apical cell membranes are shed into urine from injured podocytes: A novel phenomenon of podocyte injury. J Am Soc Nephrol. **16**, 408–416 (2005)
19. Hara, M., Yanagihara, T., Kihara, I.: Cumulative excretion of urinary podocytes reflects disease progression in IgA nephropathy and Schönlein-Henoch purpura nephritis. Clin J Am Soc Nephrol. **2**, 231–238 (2007)
20. Sato, Y., Wharram, B.L., Lee, S.K., Wickman, L., Goyal, M., Venkatareddy, M., et al.: Urine podocyte mRNAs mark progression of renal disease. J Am Soc Nephrol. **20**, 1041–1052 (2009)
21. Nakamura, T., Ushiyama, C., Suzuki, S., Hara, M., Shimada, N., Sekizuka, K., et al.: Urinary podocytes for the assessment of disease activity in lupus nephritis. Am J Med Sci. **320**, 112–116 (2000)
22. Mansur, J.B., Sabino, A.R., Nishida, S.K., Kirsztajn, G.M.: Is there a role for urinary podocyte excretion assessment in lupus nephritis? Ren Fail. **38**(4), 643–647 (2016)
23. Hara, M., Yanagihara, T., Takada, T., et al.: Podocalyxin on the glomerular epithelial cells is preserved well in various glomerular diseases. Nephron. **67**, 123–124 (1994)
24. Xu, L., Yang, H.-C., Hao, C.-M., Lin, S.-T., Gu, Y., Ma, J.: Podocyte number predicts progression of proteinuria in IgA nephropathy. Modern Pathology. **23**, 1241–1250 (2010)
25. Morioka, Y., Koike, H., Ikezumi, Y., Ito, Y., Oyanagi, A., Gejyo, F., Shimizu, F., Kawachi, H.: Podocyte exacerbate mesangial proliferative glomerulonephritis. Kidney Int. **60(6)**, 2192–2204 (2001)
26. Yu, D., Petermann, A., Kunter, U., Rong, S., Shankland, S.J., Floege, J.: Urinary podocyte loss is a more specific marker of ongoing glomerular damage than proteinuria. J Am Soc Nephrol. **16**, 1733–1741 (2005)
27. Tryggvason, K.: Unraveling the mechanisms of glomerular ultrafiltration: Nephrin a key component of the slit diaphragm. J Am Soc Nephrol. **10**, 2440–2445 (1999)
28. Patari, A., Forsblom, C., Havana, M., et al.: Nephrinuria in diabetic nephropathy of type 1 diabetes. Diabetes. **52**, 2969–2974 (2003)
29. Rodewald, R., Karnovsky, M.J.: Porous substructure of the glomerular slit diaphragm in the rat and mouse. J Cell Biol. **60**, 423–433 (1974)
30. Koop, K., Eikmans, M., Baelde, H.J., Kawachi, H., de Heer, E., Paul, L.C., Bruijn, J.A.: Expression of podocyte-associated molecules in acquired human kidney diseases. J Am Soc Nephrol. **14**, 2063–2071 (2003)
31. Huber, T.B., Bensing, T.: The slit diaphragm: a signaling platform to regulate podocyte function. Curr Opin Nephrol Hypertension. **14**(3), 211–216 (2005)
32. Benzing, T.: Signaling at the slit diaphragm. J Am Soc Nephrol. **15**, 1382–1391 (2004)
33. Doublier, S., Ruotsalainen, V., Salvidio, G., Lupia, E., Biancone, L., Conaldi, P.G., Reponen, P., Tryggvason, K., Camussi, G.: Nephrin redistribution on podocytes is a potential mechanism for proteinuria in patients with primary acquired nephritic syndrome. Am J Pathol. **158(5)**, 1723–1731 (2001)
34. Luimula, P., Aaltonen, P., Ahola, H., Palmen, T., Holthofer, H.: Alternatively spliced nephrin in experimental glomerular disease of the rat. Pediatr Res. **48**, 759–762 (2000)

35. Nakatsue, T., Koike, H., Han, G.D., Suzuki, K., Miyauchi, N., Yuan, H., Salant, D.J., Gejyo, F., Shimizu, F., Kawachi, H.: Nephrin and podocin dissociate at the onset of proteinuria in experimental membranous nephropathy. Kidney Int. **67**, 2239–2253 (2005)
36. Huh, W., Kim, D.J., Kim, M.-K., Kim, Y.G., Oh, H.-Y., Ruotsalainen, V., Tryggvason, K.: Expression of nephrin in acquired human glomerular disease. Nephrol Dial Transplant. **17**, 478–484 (2002)
37. Yuan, H., Takeuchi, E., Taylor, G.A., McLaughlin, M., Brown, D., Salant, D.J.: Nephrin dissociates from actin, and its expression is reduced in early experimental membranous nephropathy. J Am Soc Nephrol. **13**, 946–956 (2002)
38. Gadliardini, E., Benigni, A., Tomasoni, S., Abbate, M., Kalluri, R., Remuzzi, G.: Targeted downregulation of extracellular nephrin in human IgA nephropathy. Am J Nephrol. **23**, 277–286 (2003)
39. Patrakka, J., Ruotsalainen, V., Ketola, I., Holmberg, C., Heikinheimo, M., Tryggvason, K., Jalanko, H.: Expression of nephrin in pediatric kidney diseases. J Am Soc Nephrol. **12**, 289–296 (2001)
40. Gadliardini, E., Benigni, A., Tomasoni, S., Abbate, M., Kalluri, R., Remuzzi, G.: Targeted downregulation of extracellular nephrin in human IgA nephropathy. Am J Nephrol. **23**, 277–286 (2003)
41. Lavoie, J.N., Hickey, E., Weber, L.A., Landry, J.: Modulation of actin microfilament dynamics and fluid phase pinocytosis by phosphorylation of heat shock protein 27. J Biol Chem. **268**(24), 210–24,214 (1993)
42. Lee, G.J., Roseman, A.M., Saibil, H.R., Vierling, E.: A small heat shock protein stably binds heat-denaturated model substrates and can maintain a substrate in a folding-complement state. EMBO J. **16**, 659–671 (1997)
43. Huot, J., Houle, F., Spitz, D.R., Landry, J.: HSP27 phosphorylation-mediated resistance against actin fragmentation and cell death induced by oxidative stress. Cancer Res. **56**, 273–279 (1996)
44. Mehlen, P., Hickey, E., Weber, L.A., Arrigo, A.P.: Large unphosphorylated aggregates as the active form of hsp27 which controls intracellular reactive oxygen species and glutathione levels and generated a protection against TNFα in NIH-3 T3-ras cells. Biochem Biophys Res Commun. **241**, 187–192 (1997)
45. Preville, X., Schultz, H., Knauf, U., Gaestel, M., Arrigo, A.P.: Analysis of the role of Hsp25 phosphorylation reveals the importance of the oligomerization state of this small heat shock protein in its protective function against TNF-α and hydrogen peroxide-induced cell death. J Cell Biochem. **69**, 436–452 (1998)
46. Topham, P.S., Haydar, S.A., Kuphal, R., et al.: Complement-mediated injury reversibly disrupts glomerular epithelial cell actin microfilaments and focal adhesion. Kidney Int. **55**, 1763–1775 (1999)
47. Saran, A.M., Yuan, H., Takeuchi, E., McLaughlin, M., Salant, D.J.: Complement mediates nephrin redistribution and actin dissociation in experimental membranous nephropathy. Kidney Int. **64**, 2072–2078 (2003)
48. Mehlen, P., Hickey, E., Weber, L.A., et al.: Large unphosphorylated aggregates as the active form of hsp27 which controls intracellular reactive oxygen species and glutathione levels and generated a protection against TNFα in NIH-3 T3-ras cells. Biochem Biophys Res Commun. **241**, 187–192 (1997)
49. Smoyer, W.E., Gupta, A., Mundel, P., Ballew, J.D., Welsh, M.J.: Altered expression of glomerular heat shock protein 27 in experimental nephrotic syndrome. J Clin Invest. **97**, 2697–2704 (1996)
50. Smoyer, W.E., Ranson, R.F.: Hsp27 regulates podocyte cytoskeletal changes in an in vitro model of podocyte process retraction. FASEB J. **16**, 316–326 (2002)
51. Tsagalis, G.C., Nikolopoulou, N., Sotsiou, F., Hadjiconstantinou, V.: The Expression of heat shock proteins 27 and 70 in lupus nephritis. Hospital Chronicles. **1**(**3**), 125–129 (2006)
52. Eremina, V., Baelde, H.J., Quaggin, S.E.: Role of the VEGF—a signaling pathway in the glomerulus: evidence for crosstalk between components of the glomerular filtration barrier. Nephron. Physiology. **106**(2), 32–37 (2007)
53. Sugimoto, H., Hamanog, Y., Charytan, D., et al.: Neutralization of circulating vascular endothelial growth factor (VEGF) by anti-VEGF antibodies and soluble VEGF receptor 1 (sFlt1) induces proteinuria. Journal of Biological Chemistry. **278**(15), 12605–12608 (2003)
54. Yang, J.C., Haworth, L., Sherry, R.M., et al.: A randomized trial of bevacizumab, an anti-vascular endothelial growth factor antibody, for metastatic renal cancer. New England Journal of Medicine. **349**(5), 427–434 (2003)
55. Kabbinavar, F., Hurwitz, H.I., Fehrenbacheretal, L.: Phase II, randomized trial comparing bevacizumab plus fluorouracil (FU)/leucovorin (LV) with FU/LV alone in patients with metastatic colorectal cancer. Journal of Clinical Oncology. **21**(**1**), 60–65 (2003)
56. Eremina, V., Jefferson, J.A., Kowalewska, J., et al.: VEGF inhibition and renal thrombotic microangiopathy. New England Journal of Medicine. **358**(**11**), 1129–1136 (2008)
57. Patel, T., Morgan, J., Demetri, G., et al.: A preeclampsia-like syndrome characterized by reversible hypertension and proteinuria induced by the multitargeted kinase inhibitors sunitinib and sorafenib. Journal of the National Cancer Institute. **100**(**4**), 282–284 (2008)
58. Maynard, S.E., Min, J.Y., Merchan, J., et al.: Excess placental soluble fms-like tyrosine kinase 1 (sFlt1) may contribute to endothelial dysfunction hypertension, and proteinuria in preeclampsia. Journal of Clinical Investigation. **111**(**5**), 649–658 (2003)
59. Muller-Deile, J., Schiffer, M.: Renal involvement in preeclampsia: similarities to VEGF ablation therapy. Journal of Pregnancy. **2011**(176), 973 (2011)

60. Simon, M., Grone, H.J., Johren, O., et al.: Expression of vascular endothelial growth factor and its receptors in human renal ontogenesis and in adult kidney. American Journal of Physiology. **268**(**2**), 240–250 (1995)
61. Horita, Y., Miyazaki, M., Koji, T., et al.: Expression of vascular endothelial growth factor and its receptors in rats with protein-overload nephrosis. Nephrology Dialysis Transplantation. **13**(**10**), 2519–2528 (1998)
62. Matsumoto, K., Kanmatsuse, K.: Elevated vascular endothelial growth factor levels in the urine of patients with minimal-change nephrotic syndrome. Clinical Nephrology. **55**(**4**), 269–274 (2001)
63. Bailey, E., Bottomley, M.J., Westwell, S., et al.: Vascular endothelial growth factor mRNA expression in minimal change, membranous, and diabetic nephropathy demonstrated by non-isotopic in situ hybridization. Journal of Clinical Pathology. **52**, 735–738 (1999)
64. Schrijvers, B.F., Flyvbjerg, A., De Vriese, A.S.: The role of vascular endothelial growth factor (VEGF) in renal pathophysiology. Kidney International. **65**(**6**), 2003–2017 (2004)
65. Bailey, E., Bottomley, M.J., Westwell, S., et al.: Vascular endothelial growth factor mRNA expression in minimal change, membranous, and diabetic nephropathy demonstrated by non-isotopic in situ hybridization. Journal of Clinical Pathology. **52**, 735–738 (1999)
66. Kubo, A., Nishitani, Y., Minamino, N., et al.: Adrenomedullin gene transcription is decreased in peripheral blood mononuclear cells of patients with IgA nephropathy. Nephron. **85**(**3**), 201–206 (2000)
67. Yuan, H.T., Tipping, P.G., Li, X.Z., Long, D.A., Woolf, A.S.: Angiopoietin correlates with glomerular capillary loss in anti-glomerular basement membrane glomerulonephritis. Kidney International. **61**(**6**), 2078–2089 (2002)
68. Shulman, K., et al.: Expression of vascular permeability factor (VPF/VEGF) is altered in many glomerular disease. J Am Soc Nephrol. **7**, 661–666 (1996)
69. Garovic, V.D., Wagner, S.J., Petrovic, L.M., Gray, C.E., Hall, P., Sugimoto, H., Kalluri, R., Grande, J.P.: Glomerular expression of nephrin and synaptopodin but not podocin is decreased in kidney sections from women with preeclampsia. Nephrol Dial Transplant. **22**, 1136–1143 (2007)
70. Hara, A., Wada, T., Furuchi, K., Sakai, N., Kawachi, H., Shimizu, F., Shibuya, M., Matsushima, K., Yokoyama, H., Egashira, K., Kaneko, S.: Blokade of VEGF accelerates proteinuria via decrease in nephrin expression in rat crescentic glomerulonephritis. Kidney Int. **69**(**11**), 1986–1995 (2006)
71. Kriz, W.: Progressive renal failure—Inability of podocytes to replicate and the consequences for development of glomerulosclerosis. Nephrol Dial Transplant. **11**, 1738–1742 (1996)
72. Lemley, K.V., Lafayette, A., Safai, G., Derby, G., Blouch, K., Squarer, A., Myers, B.D.: Podocytopenia and disease severity in IgA nephropathy. Kidney Int. **61**, 1475–1485 (2002)
73. Kriz, W., Greitz, N., Lemley, K.V.: Progression of glomerular diseases: is the podocyte the culprit? Kidney Int. **54**, 687–697 (1998)
74. Schiffer, M., Bitzer, M., Roberts, I.S.: Apoptosis in podocytes induced by TGF-beta and Smad7. J Clin Invest. **108**, 807–816 (2001)
75. Kim, Y.H., Goyal, M., Kurnit, D.: Podocyte depletion and glomerulosclerosis have a direct relationship in the PAN-treated rat. Kidney Int. **60**, 957–968 (2001)
76. Yamaguchi, Y., Iwano, M., Suzuki, D.: Epithelial-mesenchymal transition as a potential explanation for podocyte depletion in diabetic nephropathy. Am J Kidney Dis. **54**, 653–664 (2009)
77. Miyauchi, M., Toyoda, M., Kobayashi, K.: Hypertrophy and loss of podocytes in diabetic nephropathy. Intern Med. **48**, 1615–1620 (2009)
78. Reidy, K., Susztak, K.: Epithelial-mesenchymal transition and podocyte loss in diabetic kidney disease. Am J Kidney Dis. **54**, 590–593 (2009)
79. Petermann, A.T., Pippin, J., Durvasula, R., Pichler, R., Hiromuda, K., Monkawa, T., Couser, W.G., Shankland, S.J.: Mechanical stretch induces podocyte hypertrophy in vitro. Kidney Int. **67**, 157–166 (2005)
80. Petermann, A., Floege, J.: Podocyte damage resulting in podocyturia: a potential diagnostic marker to assess glomerular disease activity. Nephron Clin Pract.**106**(**2**), 61–66 (2007)
81. Colucci, G., Floege, J., Schena, F.P.: The urinary sediment beyond light microscopical examination. Nephrol Dial Transplant.**21**(**6**), 1482–1485 (2006)
82. Wharram, B.L., Goyal, M., Wiggins, J.E., Sanden, S.K., Hussain, S., Filipiak, W.E., Saunders, T.L., Dysko, R.C., Kohno, K., Holzman, L.B., Wiggins, R.C.: Podocyte depletion causes glomerulosclerosis: diphtheria toxin-induced podocyte depletion in rats expressing human diphtheria toxin receptor transgene. J Am Soc Nephrol.**16**(**10**), 2941–2952 (2005)
83. Ronco, P., Lelongt, B., Piedagnel, R., Chatziantoniou, C.: Matrix metalloproteinases in kidney disease progression and repair: a case of flipping the coin. Semin Nephrol. **27**, 352–362 (2007)
84. Galis, Z.S., Khatri, J.J.: Matrix metalloproteinases in vascular remodeling and atherogenesis: the good, the bad, and the ugly. Circ Res. **90**, 251–262 (2002)
85. Visse, R., Nagase, H.: Matrix metalloproteinases and tissue inhibitors of metalloproteinases: structure, function and biochemistry. Circ Res. **92**, 827–839 (2003)
86. Donnelly, R., Donnelly, J., Manning, G.: Hypertension, matrix metalloproteinases and target organ damage. J Hypertens. **21**, 1627–1630 (2003)
87. Li, Y., Kang, Y.S., Dai, C., Kiss, L.P., Wen, X., Liu, Y.: Epithelial-to-mesenchymal transition is a potential pathway leading to podocyte dysfunction and proteinuria. Am J Pathol.**172**(**2**), 299–308 (2008)
88. Cheng, S., Lovett, D.H.: Gelatinase A (MMP-2) is necessary and sufficient for renal tubular cell epithelial-mesenchymal transformation. Am J Pathol. **162**, 1937–1949 (2003)
89. Tan, T.K., Zheng, G., Hsu, T.T., Wang, Y., Lee, V.W., Tian, X., Wang, Y., Cao, Q., Wang, Y., Harris, D.C.:

Macrophage matrix metalloproteinase-9 mediates epithelial-mesenchymal transition in vitro in murine renal tubular cells. Am J Pathol. **176**, 1256–1270 (2010)
90. Cheng, S., Pollock, A.S., Mahimkar, R., Olson, J.L., Lovett, D.H.: Matrix metalloproteinase 2 and basement membrane integrity: a unifying mechanism for progressive renal injury. FASEB J. **20**, 1898–1900 (2006)
91. Dimas, G., Iliadis, F., Grekas, D.: Matrix metalloproteinases, atherosclerosis, proteinuria and kidney disease: Linkage–based approaches. Hippokratia. **17(4)**, 229–297 (2013)
92. Becher, N., Hein, M., Uldbjerg, N., Danielsen, C.C.: Balance between matrix metalloproteinases (MMP) and tissue inhibitors of metalloproteinases (TIMP) in the cervical mucus plug estimated by determination of free non-complexed TIMP. Reprod Biol Endocrinol. **6**, 45 (2008)
93. Akiyama, K., Shikata, K., Sugimoto, H., Matsuda, M., Shikata, Y., Fujimoto, N., Obata, K., Matsui, H., Makino, H.: Changes in serum concentrations of matrix metalloproteinases, tissue inhibitors of metalloproteinases and type IV collagen in patients with various types of glomerulonephritis. Res Commun Mol Pathol Pharmacol. **95**, 115–128 (1997)
94. Koide, H., Nakamura, T., Ebihara, I., Tomino, Y.: Increased mRNA expression of metalloproteinase-9 in peripheral blood monocytes from patients with immunoglobulin A nephropathy. Am J Kidney Dis. **28**, 32–39 (1996)
95. Urushihara, M., Kagami, S., Kuhara, T., Tamaki, T., Kuroda, Y.: Glomerular distribution and gelatinolytic activity of matrix metalloproteinases in human glomerulonephritis. Nephrol Dial Transplant. **17**, 1189–1196 (2002)
96. Sanders, J.S., van Goor, H., Hanemaaijer, R., Kallenberg, C.G., Stegeman, C.A.: Renal expression of matrix metalloproteinases in human ANCA-associated glomerulonephritis. Nephrol Dial Transplant. **19**, 1412–1419 (2004)
97. McMillan, J.I., Riordan, J.W., Couser, W.G., Pollock, A.S., Lovett, D.H.: Characterization of a glomerular epithelial cell metalloproteinase as matrix metalloproteinase-9 with enhanced expression in a model of membranous nephropathy. J Clin Invest. **97**, 1094–1101 (1996)
98. Keeling, J., Herrera, G.A.: Human matrix metalloproteinases: characteristics and pathologic role in altering mesangial homeostasis. Microsc Res Tech. **71**, 371–379 (2008)

On Research and Translation of Urinary Biomarkers

7

Youhe Gao

Abstract

From the theory of homeostasis, it can be deduced that urine is the source of sensitive disease markers reflecting early changes of the body. The study of urinary biomarkers using animal models is essential to prove this theory and encourage people to continue exploring the potential of urine. In clinical research, when disease-related changes are greater than individual variances, disease-related biomarkers with potential clinical application can be obtained by directly dividing samples into disease groups and control groups. To discover small early changes in disease, pre-and-post control of the same person can minimize most interfering factors. In this way, changes in urinary proteins before, during and after disease and/or treatment can be found, which can provide useful information for early detection and evaluation of the disease condition and treatment effect. In the study of clinical urinary biomarkers, regional and ethnic factors cannot be completely ignored. Diseases such as autism, which have only social behavior changes, may also be reflected in the urine proteome. Current research on urinary biomarkers is not sufficient to earn the recognition it deserves in the field of biomarkers. The recognition of urinary biomarkers will require the cooperation of more doctors and scientists and the participation of more foundations and companies.

Keywords

Homeostasis of internal environment · Animal models · Early biomarkers · Regions and nationalities · Autism · Pre-and-post control · Urinary protein preservation · Urimem

Early diagnosis is almost always better than late diagnosis, even if medical intervention sometimes needs to be started later by design. For early diagnosis, the vast majority of efforts are still devoted to developing more precise instruments and detection methods to find smaller changes in blood. As for where to find these changes, researchers in the biomarker field seem to have little doubt. Blood is the first choice. After all, blood communicates with all organs of the body and carries out direct material and energy exchange all the time. To change current thinking, we should not only try to be as sensitive and precise as possible in detection but should also present new ideas to search for biomarkers at

Y. Gao, MD/PhD (✉)
Department of Biochemistry and Molecular Biology, Beijing Key Laboratory of Gene Engineering Drug and Biotechnology, Beijing Normal University, Beijing, China
e-mail: gaoyouhe@bnu.edu.cn

H. M. Baptista Carreira dos Santos (ed.), *Translational Urinomics*, Advances in Experimental Medicine and Biology 1306, https://doi.org/10.1007/978-3-030-63908-2_7

different sites. Do the smallest changes in the human body occur earliest in blood?

Homeostasis of the internal environment was an important discovery in physiology [1, 2]. In short, as an important component of the internal environment, blood is regulated by the body's homeostatic mechanisms and is kept as stable as possible. This effect is very important to both human health and survival. However, this effect makes it very difficult to find changes in blood, especially when changes occur relatively early and are small. Small changes are gradually eliminated by the homeostatic mechanisms of the body during the process of detection. The result is that we have always wondered whether our detection methods and instruments are stable enough. However, early diagnosis is so important to medicine that it can give the body a chance to be repaired when the changes in the body are still reversible. Although late diagnosis has better reproducibility, the reversibility of changes in the body is not as good at late stages. The treatment of advanced diseases is much more difficult if they can still be treated.

A few years ago, the urine biomarker theory was proposed, which defined biomarkers as measurable changes related to diseases, emphasizing the important characteristic of biomarkers as "changes." Moreover, due to the control of the homeostatic mechanisms, early small changes in blood are eliminated, while urine collects waste from the whole body and is richer in changes than blood, making it a better biomarker source [3].

Although this theory makes sense, habitual thinking still controls the biomarker field. The acceptance of new ideas still requires much work, and it is difficult to prove that early disease changes can be reflected in urine. Many people think that the changes found in clinical samples of urine are not necessarily disease-related because many other factors can also change urine. When the sample size is not large, the persuasibility is even more limited. However, when people do not have enough confidence, who will provide the funding and which researchers will perform large clinical studies with many uncontrolled variables?

One simple and effective way to prove the urine biomarker theory is to use animal models. The advantage of using animal models is that the researcher can know the exact time when the disease starts, which is equivalent to knowing the absolute early stage of the disease. This early stage can be much earlier than the early stage observed in clinics, which is limited by the time of the chief complaint or a regular check-up, whichever comes earlier. Animal models can also eliminate interfering factors caused by medicine and other lifestyle factors, such as tobacco, alcohol, beverages, and nutrition [4].

Although the conclusions drawn from animal models may not be directly applicable in clinical practice, if urine can reflect the very early stages of diseases in animal models, it can also eventually reflect the very early stages of diseases in humans. This information has not been taken seriously in the biomarker field. It is hoped that results from animal models will arouse the attention of scientists, doctors, companies and foundations. Only sustained investment in human and material resources can lead to the translation of urinary biomarkers into clinics.

There is a large difference between animal model research and clinical research. Animal model experiments can be well designed so that samples before and after the disease process can be collected and compared. Such comparisons can more accurately reflect changes related to the occurrence and development of disease. However, in clinical research, the earliest samples that can be obtained are from when the patient first comes to visit the doctor. Healthy controls are often missing if physical examination samples are not preserved before the onset of disease. Thus, comparisons before and after disease cannot be made. In this case, only samples of other people can be chosen as the controls, but no control is better control than the patient's own sample. Clinical samples are inherently affected by various influencing factors, and it is difficult to find a perfectly matched control. Thus, it is difficult to obtain urinary changes only related to diseases through the use of limited clinical samples because of these confounding factors. Candidate urinary disease biomarkers identified in the presence of various factors are also difficult to be verified in other groups of samples. Therefore, such research is often paused due to various reasons, such as funding and human resource limitations, before the urine biomarker theory can be approved by researchers and clinicians in a convincing way.

At this stage, more laboratories are needed to perform more studies using animal models. Only when many laboratories successfully prove that urine can reflect early changes of various diseases will more resources be poured into this field, will the development of the urine marker field be accelerated, and will the clinical application of urine biomarkers begin to gradually emerge. Once this positive feedback starts, the impact of urine biomarkers on medicine can occur in many different fields very quickly.

For researchers and doctors who insist on starting clinical trials as soon as possible, it is important to fully understand the various factors that affect urine. Understanding the influence of various lifestyle factors and drug factors is helpful for designing experiments to avoid, reduce or offset the influence of such factors. There are variances among healthy individuals. If the differences caused by disease are sufficiently large and much larger than the variances among healthy individuals, then disease-related changes can be found using the method of disease-healthy grouping. However, it is difficult to anticipate how large early changes can be in certain chronic diseases. Because early changes of diseases are usually very small, their magnitude may not exceed the variance among healthy individuals. In many cases, it is difficult to find a perfectly matched control group. At this time, using one's own pre-to-post control, as in animal model experiments, is a direction for future clinical application. The self-control comparison can theoretically eliminate the impact of many different lifestyle factors on urine and can identify urinary changes that are only related to disease. Moreover, self-control is also in line with the direction of personalized medicine in the future; after all, every person has a unique genetic background and unique living environment. In this way each person's own urinary status can be accurately recorded, and changes before and after a period of time can be analyzed [5].

One of the disadvantages of the pre-and-post comparison is that the current medical system does not routinely preserve samples from a healthy people. It is difficult to find urine samples to use as a healthy self-control when people are ill, although urine can be stored economically and noninvasively. The use of urine for early disease diagnosis may not have been realized nor recognized, and the cost of the cryopreservation of urine is too high. The method of urimem solves the problem of cost. Proteins in urine can be adsorbed onto various membranes when urine is filtered, and other macromolecules, such as nucleic acids, can also be selectively adsorbed onto membranes that are sticky for nucleic acids. Through drying and vacuum preservation, most preservation costs can be saved. The urimem preservation method enables macromolecules in urine to be simply and economically stored in medical records. With the availability of these samples and biological information, the diagnosis and monitoring of disease progression will become very different from current practice and the face of medicine will change [6].

In short, research on animal models is still of great significance at this stage, and more laboratory involvement is needed. Without considering all the factors that influence clinical research, blindly starting clinical research will hardly produce convincing results and will lead to erroneous conclusions that urinary biomarkers are unstable, unable to distinguish different disease states, and have no future. This is the biggest mistake that can be made at this stage for the urinary biomarker field.

Compared with animal model research, clinical research has several common problems. One problem is that in an animal model study there is often one subtype of disease involved, and it is easier to obtain consistent results when studying the same subtype of animals. However, in clinical studies, in the disease group, only clinical manifestations and/or some laboratory tests are similar, and individuals may not suffer from exactly the same disease. Another factor that is difficult to control is the time point, that is, the specific stage of the disease. We used to naively hope that a disease can be reflected by one marker that gradually increases with the aggravation of the disease and decreases with the alleviation of the disease. In fact, diseases may be qualitatively different in different stages, and the combination of biomarkers may be essentially different in different stages of the same disease. Grouping different subtypes of diseases into one group and grouping diseases at different time stages into one group makes it difficult to obtain reproducible and verifiable results in biomarker research. The specificity of biomarkers is also a problem with this "one disease one biomarker" ideal. Only through exhaustive studies will complicated combinations of biomarkers become so-called practically usable specific biomarkers.

Many of the studies described in this chapter also fail to take into account the above mentioned differences in disease processes and subtypes due to their preliminary nature.

In fact, at this stage, many factors that need to be considered have not been considered yet, and sometimes, they are avoided purely by chance. However, these factors may need to be considered in future research. For example, people from one region may have different urinary protein profiles than those from other regions due to various reasons. First, it should be stressed that the scope of the research mentioned below is not at a very large scale. The specific results of this research should be viewed with caution, but it is still worth knowing that such differences may exist. One study showed human urinary proteome differences in samples from people from three different regions (Haikou, Xi 'an and Xining). The study conducted nonlabeled, quantitative analysis of total urinary proteins from 9 individuals in each region. Among the total 1898 proteins identified, 16 differentially expressed proteins in males and 84 in females were significantly different ($P < 0.05$) from different regions [7]. At this stage, the cost of increasing the scale and improving the reliability is relatively high, especially when the use of mass spectrometers is still relatively expensive. However, the results of are still worth reporting. Please pay attention to the possibility of regional influences on urinary proteins when urine samples from different regions are used for clinical research. After the article was published and before writing this chapter, we reanalyzed the differences via exhaustive random grouping. The samples from the two locations were randomly divided into two equal groups. The numbers of differentially expressed proteins were calculated for all possible random groupings to evaluate the possibility of random generation of urine differential proteins from the two sites. The differences that we regarded as regional differences were not likely to be generated randomly. Of course, not all locations may show differences in urine proteomes. These results only suggest that there may be regional differences worth noting when performing multicenter clinical urinary biomarker studies in the future.

In addition to the influence of different regions, in a casual study, we found that different ethnic groups in the same region also had differences in urinary protein profiles. The Han and Li nationalities in Hainan have lived in the same area for thousands of years, and genetic research has shown that they have some specific genetic characteristics. Due to various reasons, such as personnel and instruments, the number of samples identified by nonlabeled quantitative proteomics at that time was very limited and only included the Han and Li nationalities, with 3 males and 3 females each, totaling 12 samples. Because of the small amount of data and the long-term delay of analysis, the study was considered to hardly yield convincing or interesting results. In an unsupervised cluster analysis without any positive expectation, we unexpectedly observed that without any data processing, these 12 samples actually divided themselves into four groups, namely, Han men, Han women, Li women and Li men. If men and women were divided into two groups according to nationality, 53 differentially expressed proteins were produced, but if the 12 samples were randomly and equally divided into two groups, all the grouping possibilities were taken into account, and the average number of differentially expressed proteins was only 12. In other words, there may still be large differences in urine proteins among these two nationalities, even though the two groups have lived together for many years and even if the differences in their lifestyle are no longer very significant. This very preliminary result makes us interested in continuing to analyze the results because unsupervised clustering does not look for differential proteins first and then display differentially expressed proteins as an aesthetically pleasing map. Instead, unsupervised clustering refers to clustering all identified proteins without any pretreatment. If the categories can be classified under the condition of unsupervised clustering, the differences between categories are considered relatively obvious. Of the 1555 total urine proteins identified, 25 had significant differences, 16 of which have been reported to be biomarkers of diseases. This preliminary result at least suggests that nationality may be a factor worth considering in future clinical studies [8].

The results mentioned above are from very small sample sizes. The results are not recommended as being conclusive but rather are more suitable as a clue. Laboratories with energy and resources should consider using a larger scale and more rigorous design to verify these results.

If there is a large difference among people, can urinary biomarkers ever be translated into clinical use?

It can be assumed that if the difference caused by a disease is larger and does not coincide with that of healthy people, then it is still possible to group samples into disease and control groups. Thus, which diseases are likely to be studied in this way? The answer is difficult to guess accurately.

Due to the limited understanding of the field, the following is only a brief introduction of what we have tried with collaborators.

First, we discuss a preliminary result from autism research. Autism is a complex neurological disease. The cause of autism is not clear, and there are many contributing factors, including many genetic factors. Moreover, the main manifestation of autism is a social abnormality, and there is no clear organic manifestation in the body. Is it possible to find differences between autistic and healthy control groups? I suspect that infectious diseases may cause greater differences compared to healthy controls and have significant clinical manifestations.

The following experiment is a very casual preliminary attempt performed without any expectations. Eighteen urine samples of 3–7-year-old autistic patients were voluntarily donated from autistic children at the Sunshine Angel Special Training Center of Fengtai District in Beijing, which specializes in the rehabilitation of children diagnosed with autism. Urine of healthy children came from children of the same age of the staff of the College of Life Sciences, Beijing Normal University. The urine of 18 autistic children aged 3–7 years old and of 6 healthy children was analyzed by data-independent quantitative proteomics. Compared to healthy children, 120 differential proteins were identified in the urine of autistic children (Fold Change $\geq$1.5 or $\leq$0.67 and $P < 0.01$). Twenty-four of these proteins have been reported to be associated with autism. When we observed these differences, we still doubted whether they were generated by random chance. Is it possible that any 24 samples with 1770 proteins identified will produce some number of differential proteins if they are randomly divided into two groups of 18 to 6? We tried all the combinations using a simple homemade program, and the average number of differential proteins was 10. That is, the urine proteomes of 24 children were randomly divided into two groups (18 to 6) with 134,596 grouping possibilities, and an average of 10 differential proteins were identified using the exact same threshold used above. This finding means that approximately 10 of the 120 differential proteins might be randomly generated and have nothing to do with the real difference between autistic and healthy children. Therefore, approximately 110 differential proteins are truly related to differences between autistic and healthy children [9]. It was unexpected to find such a large difference in such a small sample size. Perhaps this difference is because of the following factors: children's lifestyles are relatively simple in that they do not smoke or drink; the patients happened to have the similar or even same subtype of autism; and the patients happened to be at the same stage of the disease. Even so, it was unexpected to see such statistically significant differences.

In addition to statistical support, these differential proteins were found to participate in multiple signaling pathways, of which endogenous cannabinoid developmental neural pathways, synaptic long-term inhibition pathways, synaptic signaling pathways and mTOR signaling pathways have been reported to be associated with autism. In addition, some pathways that may be related to autism were also significantly enriched, such as the Ephrin receptor signaling pathway, CREB signaling pathway and STAT3 pathway in neurons. These results makes us believe that the differential proteins in urine may actually reflect differences of autistic children. These results provide objective indicators for the clinical diagnosis of autism and also provide clues for understanding the pathological mechanisms and opportunities for early intervention.

If a disease with only social symptoms can be distinguished using the urine proteome so early, what other diseases should not be tried?

If a disease has significant physical signs, should it be easier to distinguish using urine? Urine samples of three different subtypes of abdominal anaphylactoid purpura according to traditional Chinese medicine were collected from our partner Beijing Children's Hospital. The treatments for different subtypes of this disease are different in China, but there is no objective differential diagnostic biomarker. In short, the clinical manifestations of the three subtypes are somewhat different and cannot be clearly distinguished. In total, 59 patients were divided into 3 subtypes and a healthy control group. In the analysis of these samples, 75 differentially expressed proteins could be used to distinguish healthy people from abdominal anaphylactoid purpura. Ten differentially expressed protein panels were selected to distinguish these three subtypes from each other [10]. Because abdominal anaphylactoid purpura is a children's disease, the influence of lifestyle may be relatively small, and different clinical manifestations indicate significant changes of the affected organs. It was not surprising that he pathological changes of different organs were reflected in urinary proteins. Similar diseases should be studied the same way. Perhaps this is the earliest result that can be translated into clinical practice.

Some researchers prefer to use clinical samples directly for biomarker discovery instead of verifying clues found in animal model studies. After all, diseases in animal models are different from those in patients, and the complexity between animal models and patients varies greatly. However, how can we find clues in relatively small numbers of clinical samples? One attempt was to find tumor-related differential proteins by comparing the urine proteome before and after surgery among brain tumor patients in Peking University International Hospital. By comparing the urine proteome of 5 pairs of glioma patients before and after surgery, 27 proteins were found to have significant changes. Many of these proteins have been reported by previous studies in other laboratories to be related to glioma. The biological processes represented by these differential proteins were related to glioma occurrence and development, autophagy and angiogenesis. This kind of pre-post control study is more suitable for offsetting various known or even unknown influencing factors. This study sought to find clues by comparing urine before and after surgery. For various reasons, only five pairs of samples of these five patients were evaluated. Even so, there were still some consistent changes, and it seems that these patients may still have certain similarities [11].

I think, for urinary biomarkers, the closest way to translate them into clinics is using markers for pre-post comparisons of the same person. In this way, most confounding factors can be offset as long as there are no significant changes in lifestyle between the two states being compared. Moreover, such research does not require studying many cases at the same time for statistical purposes. After all, studying a large number of people requires more resources. The clinical samples required for a large study will not be available according to the requirements of the research, and it will take time to collect samples. We hope the importance of using the pre-post control for early diagnosis will be realized and recognized and that the collection of urine samples from healthy people as future controls can be started. However, not enough attention has been paid to date, although there is a very simple and economical way to preserve urinary proteins on a membrane, urimem. Overall, the amount of money needed to preserve urine should be negligible compared to the increasing cost of disease treatment if a diagnosis is made late. Due to insufficient attention, it is still very difficult to obtain urinary protein samples when people are healthy. To compare urine samples of the two time points before and after, it is necessary to design a scenario that is helpful for clinical application in the near future. One clinical scenario is drug therapy. Urine samples before, during and after lung cancer treatment were obtained from Peking Union Medical College Hospital after obtaining informed consent and passing ethical approval without affecting patients' treatment. By comparing the urine proteome of the same patient before, during and after treatment, differentially expressed urinary proteins were identified. The related biological processes were then provided to the clinician to see whether the changes in urine could provide a picture of the drug effects on the patient. This strategy may be helpful for guiding future clinical medication.

In the analysis of several samples before, during and after drug treatment of 8 patients, the only preliminary result that we could obtain was that the urine proteomes before and after the treatment were different. This is because tumors from different patients are different, and the stages and treatments are different too. Through the analysis of these differential proteins, we could obtain some hints, such as which biological processes and which pathways had changed. When the clinicians saw these results, they felt that they reflected the overall clinical manifestations of the patients. The paper is still in the preprint stage [12]. We do not yet know whether this type of article is publishable in current peer review system. We hope that this form of pre-post treatment research can help clinicians find correlations between the biological process shown in urine and the clinical effects of a drug in the future. Perhaps this type of research can help clinicians continue to insist on or change a treatment over time. At present, these ideas are still in the stage of proof of concept. It is still difficult to conclude which tumors, which stages of patients, which drugs to treat, and which stable proteome changes are related through these small numbers of examples. More doctors and scientists are needed to use this strategy to eventually produce a large database of urinary protein changes with clinical manifestations. With a larger and more detailed database, we may find similar cases more easily and make medical decisions more accurately.

The current research is far from sufficient in terms of both quantity and quality. More cooperation between doctors and scientists interested in urinary biomarkers and more participation of foundations and companies are needed. Only sufficient high-quality data can be used to identify better correlations between urinary protein changes and the efficacy of disease management.

References

1. Cannon, W.B.: Physiological regulation of normal states: some tentative postulates concerning biological homeostatics. In: Ses Amis, ses Colleges, ses Eleves. Les Éditions Médicales, Paris (1926)
2. Cannon, W.B.: The Wisdom of the Body. Norton, New York (1939)
3. Gao, Y.: Urine—an untapped goldmine for biomarker discovery? Sci. China Life Sci. **56**(12), 1145 (2013)
4. Gao, Y.: Roadmap to the urine biomarker era. MOJ Proteomics Bioinform. **1**(1), 00005 (2014)
5. Gao, Y.: Personal standard for personalized medicine. Author J. **2**(2), 00020 (2017)
6. Jia, L., Liu, X., Liu, L., Li, M., Gao, Y.: Urimem, a membrane that can store urinary proteins simply and economically, makes the large-scale storage of clinical samples possible. Sci. China Life Sci. **57**(3), 336–339 (2014)
7. Wu, J., Qin, W., Pan, L., Wang, X., Zhang, B., Shan, G., Gao, Y.: Regional differences of the urinary proteomes in healthy Chinese individuals. Chinese Medical Sci J. **34**(3), 157–167 (2019). https://doi.org/10.1101/190710
8. Zhang, F., Li, X., Ni, Y., Shan, G., Gao, Y.: Preliminary study of the urinary proteome in Li and Han ethnic individuals from Hainan. Sci China Life Sci. (2020). https://doi.org/10.1007/s11427-018-9485-4

9. Meng, W., Huan, Y., Gao, Y.: Urine proteome in children with autism. http://www.paper.edu.cn/releasepaper/content/202001-124 (2020)
10. Jia, L., Wu, J., Wei, J., Du, L., Wang, P., Zhang, Y., Yu, Y., Wang, X., Yang, Y., Gao, Y.: Proteomic analysis of urine reveals biomarkers for the diagnosis and phenotyping of abdominal-type Henoch-Schonlein purpura. Translational Pediatrics In print
11. Wu, J., Zhang, J., Wei, J., Zhao, Y., Gao, Y.: Urinary biomarker discovery in gliomas using mass spectrometry-based clinical proteomics. Chin Neurosurg J. **6**, 11 (2020)
12. Zhang, Y., Heng, Z., Wang, Y., Gao, Y.: Changes of urine proteome in patients with advanced lung cancer after different drug treatment. http://www.paper.edu.cn/releasepaper/content/202003-251 (2020)

8 A Mechanistic-Based and Non-invasive Approach to Quantify the Capability of Kidney to Detoxify Cysteine-Disulfides

Clara Gonçalves-Dias, Catarina O. Sequeira, João B. Vicente, M. João Correia, Nuno R. Coelho, Judit Morello, Alexandra M. M. Antunes, Karina Soto, Emília C. Monteiro, and Sofia A. Pereira

Abstract

Our general goal was to non-invasively evaluate kidney tubular dysfunction. We developed a strategy based on cysteine (Cys) disulfide stress mechanism that underlies kidney dysfunction. There is scarce information regarding the fate of Cys-disulfides (CysSSX), but evidence shows they might be detoxified in proximal tubular cells by the action of *N*-acetyltransferase 8 (NAT8). This enzyme promotes the addition of an *N*-acetyl moiety to cysteine-*S*-conjugates, forming mercapturates that are eliminated in urine. Therefore, we developed a strategy to quantify mercapturates of CysSSX in urine as surrogate of disulfide stress and NAT8 activity in kidney tubular cells. We use a reduction agent for the selective reduction of disulfide bonds. The obtained *N*-acetylcysteine moiety of the mercapturates from cysteine disulfides was monitored by fluorescence detection. The method was applied to urine from mice and rat as well as individuals with healthy kidney and kidney disease.

Keywords

Disulfide stress · *N*-acetyltransferase-8 · Cysteine · Mercapturate pathway · Uremic toxins · Mercapturic acid · Kidney function marker · Proximal tubular cell

C. Gonçalves-Dias · C. O. Sequeira · M. J. Correia · N. R. Coelho · J. Morello · E. C. Monteiro
S. A. Pereira (✉)
CEDOC, Chronic Diseases Research Centre, NOVA Medical School|Faculdade de Ciências Médicas, Universidade NOVA de Lisboa, Lisbon, Portugal
e-mail: catarina.sequeira@nms.unl.pt; mjoao.correia@nms.unl.pt; judit.morello@nms.unl.pt; emilia.monteiro@nms.unl.pt; sofia.pereira@nms.unl.pt

J. B. Vicente
Instituto de Tecnologia Química e Biológica, Universidade Nova de Lisboa, Oeiras, Portugal
e-mail: jvicente@itqb.unl.pt

A. M. M. Antunes
Centro de Química Estrutural (CQE), Instituto Superior Técnico, Universidade de Lisboa, Lisbon, Portugal
e-mail: alexandra.antunes@tecnico.ulisboa.pt

K. Soto
CEDOC, Chronic Diseases Research Centre, NOVA Medical School|Faculdade de Ciências Médicas, Universidade NOVA de Lisboa, Lisbon, Portugal

Hospital Prof. Dr. Fernando da Fonseca, EPE, Amadora, Portugal

H. M. Baptista Carreira dos Santos (ed.), *Translational Urinomics*, Advances in Experimental Medicine and Biology 1306, https://doi.org/10.1007/978-3-030-63908-2_8

8.1 Introduction

Cysteine-disulfides (CysSSX) include cystine (two cysteines linked via a disulfide bridge, CysSSCys) and mixed cysteine disulfides (disulfide bond between cysteine (Cys) and another thiol including glutathione (GSH) (CysSSG), Cys (CysSSCys), coenzyme A (CoA) (CysSSCoA), sulfite (SulfoCys) or homocysteine (HCys) (CysSSHCys) [1] (Fig. 8.1). We have previously shown an increase in the formation of CysSSX kidney using an animal model of hypertension [2], that is related to increased oxidative and inflammatory status [3]. The increased formation of CysSSX and other low molecular weight (LMW) disulfides has also been described as a mechanism of redox signaling associated with acute organ inflammation events [4]. This mechanism was named disulfide stress and might represent a new paradigm in inflammatory/oxidative conditions, as it is independent of the classical view of GSH redox status. Cys is less likely to remain reduced in an oxidative environment than GSH [5, 6]. Additionally, the reaction of Cys with reactive compounds forming Cys-conjugates has been reported as far more stable and faster than the reaction with GSH [7]. On the other hand, Cys is highly abundant in kidney [8, 9] which is a primordial organ for oxidation due to its high metabolic rate and strong dependence on oxidative phosphorylation to produce energy [10]. Even though, scarce information is currently available in the literature regarding the fate of CysSSX. Evidence shows that, similarly to the cysteine-*S*-conjugates which are not disulfides (CysSX), CysSSX might be detoxified by the action of *N*-acetyltransferase 8 (NAT8) [1, 11, 12] (Fig. 8.1). NAT8 has been associated to the control of blood pressure and regulation of kidney function [13–15] and it is the last enzyme of the mercapturate pathway [12]. This pathway mostly takes place in kidney tubular cells [16] and it is NAT8 that renders cysteine-*S*-conjugates prone to urinary elimination by promoting the addition of an *N*-acetyl moiety to the cysteine-*S*-conjugates (Fig. 8.1) [1, 14].

The mechanisms beyond LMW disulfides that contribute for disulfide stress signaling and effects are still to be unveiled, but it might contribute to an adaptive response to an acute insult [2, 4]. Our general aim was to find out a strategy to non-invasively measure tubular disulfide stress in kidney. Therefore, we developed a method to quantify mercapturates of CysSSX (*N*-acetyl-CysSSX) in urine samples (Fig. 8.1) and we used it to evaluate the capability of kidney in detoxifying CysSSX in volunteers, patients after an episode of acute kidney injury (AKI), mice and rats.

8.2 Experimental

8.2.1 Method Development

Our strategy aiming to evaluate the capability of kidney to detoxify CysSSX and address renal disulfide stress was based on three basic premises: (a) insult-triggered disulfide stress in kidney increases CysSSX formation [2, 4], (b) cysteine-*S*-conjugates undergo NAT8-mediated *N*-acetylation [11, 12] and are eliminated in urine as mercapturates [1, 14] and (c) NAT8 is mostly expressed in kidney [16] and is related to kidney function [13, 15]. Herein, we developed a method to quantify urinary *N*-acetyl-CysSSX (Fig. 8.1).

For that, we used a reducing agent, tris(2-carboxyethylphosphine) (TCEP), that allows the selective reduction of disulfide bonds (converting the oxidized form to the reduced thiol, i.e. *N*-acetylcysteine) (Fig. 8.1) [18]. We named the *N*-acetylcysteine moiety resultant from the reduction of urinary *N*-acetyl-CysSSX of uNAC and we use it as a surrogate of *N*-acetyl-CysSSX. (Fig. 8.1). Then, by using a selective fluorescent reagent for thiol groups, 7-fluorobenzofurazan-4-sulfonic acid ammonium salt (SBD-F) [19], uNAC can be monitored through fluorescence detection (FD), providing high sensitivity to the method, similarly to what has been described for other thiols [20]. The coupling with a chromatographic procedure allows the separation of uNAC from the other endogenous thiols like cysteinylglycine (CysGly), GSH, Cys, CoA, cysteamine and Hcys (Sects. 8.2.2.3 and 8.2.2.4).

We have evaluated whether the method is able to distinguish uNAC from *N*-acetylcysteine

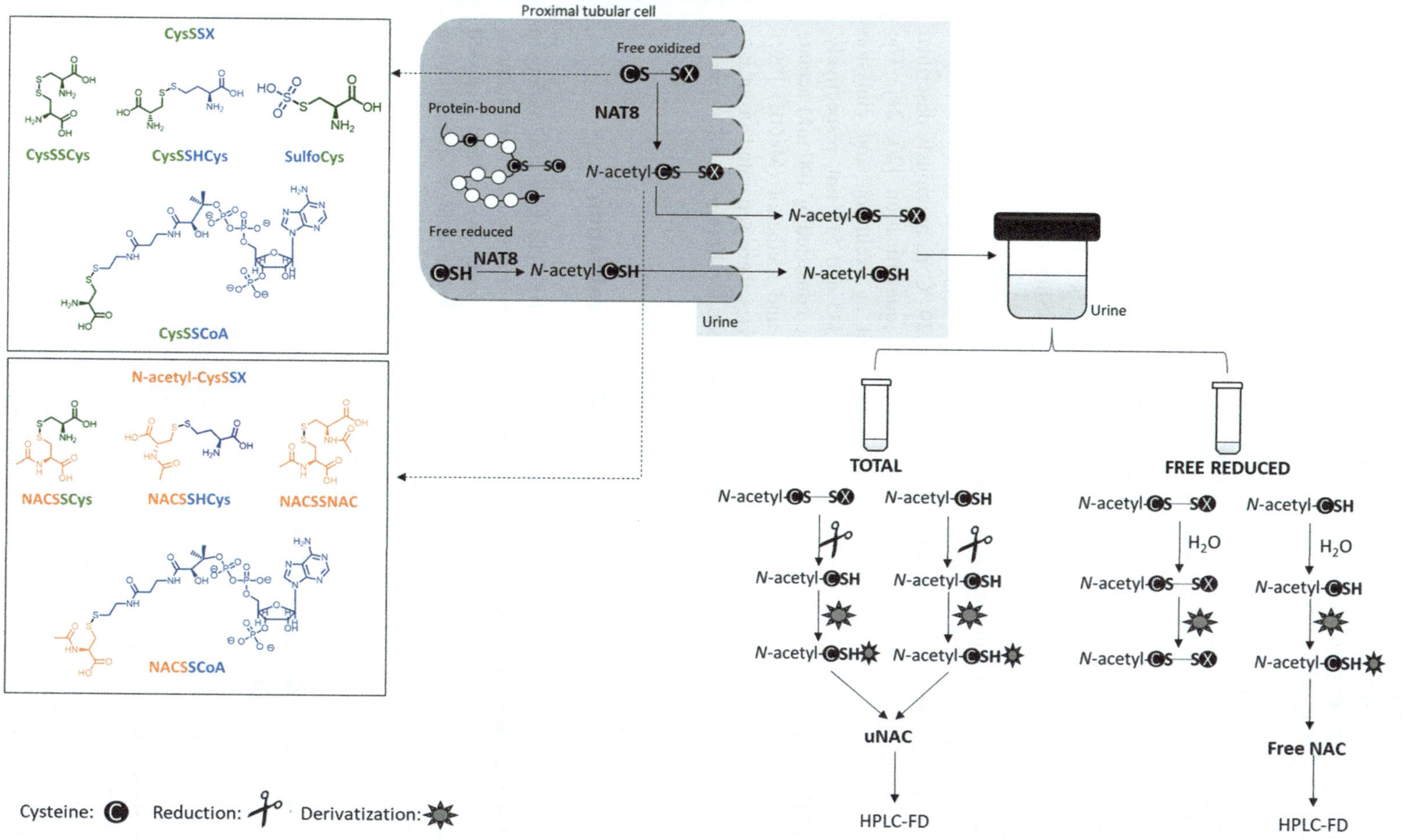

Fig. 8.1 Rationale for method development. Cysteine-disulfides are detoxified by N-acetyltransferase 8 forming mercapturates that are eliminated in urine. The developed method is able to quantify urinary mercapturates of cysteine-disulfides, using uNAC as a surrogate. The method takes advantage of a reduction agent that allows the selective reduction of disulfide bonds followed by derivatization with a selective fluorescent reagent for thiol groups. uNAC can then be monitored through fluorescence detection in a high-performance liquid chromatography system. To obtain free reduced NAC, the same protocol is followed although there is no addition of the reduction agent, allowing the derivatization of only the naturally free reduced NAC. CoA, coenzyme A; Cys, cysteine; CysSSX, cysteine-disulfides; HCys, homocysteine; HPLC-FD, high-performance liquid chromatography with fluorescence detection; NAC, N-acetylcysteine; N-acetyl-CysSSX, mercapturates of cysteine-disulfides; NAT8, N-acetyltransferase 8; Sulfo, sulfite; uNAC, urinary surrogate of mercapturates from cysteine-disulfides

(NAC) freely existing in urine (Sect. 8.2.3.7). Moreover, we ascertain that the uNAC that is being quantified is a surrogate of *N*-acetyl-CysSSX that is excreted from tubular cells and not from other cells further undergoing glomerular filtration. For this, we quantified the levels of uNAC in serum samples from volunteers, following the protocols described in Sects. 8.2.2.3 and 8.2.2.4 (Fig. 8.1). The levels of free NAC were also quantified, using for this the protocol described in Sect. 8.2.3.7. (Fig. 8.1). As so, if uNAC formed at the hepatocytes would be excreted in bile the levels of uNAC would not be detectable in serum samples.

Therefore, serum samples were collected from volunteers with normal kidney function, as shown by glomerular filtration rate superior to 90 mL/min/1.73m^2.

8.2.2 Method Description

8.2.2.1 Chemicals

High-performance liquid chromatography (HPLC)-grade solvents were purchased from VWR (Belgium). All reagents were purchased from Sigma-Aldrich (USA), except for trichloroacetic acid (TCA), which was purchased from Roth (Germany).

8.2.2.2 Stock and Calibration Solutions

Calibration solutions (CS) were prepared from successive dilutions in phosphate buffered saline (PBS 1×) to obtain eight different concentrations of NAC, namely, 2.2, 12.5, 18.8, 25, 150, 400 and 600 μM.

8.2.2.3 Reduction and Derivatization

An initial volume of 50 μL of each CS, quality control solution (QC) or sample were treated with TCEP (100 g/L, 5 μL) in deionized water for reduction of disulfide bonds. After an incubation period of 30 min at room temperature (RT), proteins were precipitated with TCA (100 g/L) in 1 mM ethylenediaminetetraacetic acid (EDTA) (45 μL)-containing deionized water. The mixture was then centrifuged (13,000 *g*, 10 min, 4 °C) and the supernatant collected into a new lab tube containing 1.55 M sodium hydroxide (NaOH) (5 μL), 125 mM sodium tetraborate buffer ($Na_2B_4O_7$, pH 9.5) with 4 mM EDTA (62.5 μL) and 25 μL of SBD-F (1 g/L) in 125 mM $Na_2B_4O_7$ buffer with 4 mM EDTA. The final mixture was vortexed and incubated in the dark, at 60 °C for 1 h, to complete derivatization of free sulfhydryl groups.

8.2.2.4 Chromatographic Conditions

A volume of 10 μL of the solution prepared as described in Sect. 8.2.2.3 was analyzed by HPLC-FD on a Shimadzu LC-10 AD VP (Shimadzu Scientific Instruments Inc) system, using a reversed-phase C18 LiChroCART 250 × 4 column (LiChrospher 100 RP-18, 5 μm, VWR, USA), at 29 °C, as previously described for other thiols quantification [20, 21]. The detector was set at excitation and emission wavelengths of 385 and 515 nm, respectively. The mobile phase consisted of 100 mM acetate buffer (pH 4.5) and methanol (MeOH) [99:1 (v/v)]. The analytes were separated in an isocratic elution mode for 20 min, at a flow rate of 0.8 mL/min.

8.2.3 Method Validation

The validation criteria were defined according to standard procedures for bioanalytical methods [22]. For this purpose, the criteria assessed included linearity, lower limit of quantification (LLOQ), higher limit of quantification (HLOQ), and four quality control (QC) samples were prepared to obtain concentrations independently from CS. Selectivity, carry-over effect, accuracy, intra- and inter-assay precision and stability after freezing cycles were studied.

8.2.3.1 Selectivity and Carry-over Effect

Blank samples of PBS 1×, MeOH or mobile phase were prepared as previously described (Sect. 8.2.2.3). The carry-over effect was studied by preparing and injecting into the HPLC system three blank samples of PBS 1× after the analysis of the highest CS sample.

8.2.3.2 Linearity

Three calibration curves were prepared from different stock solutions. Each calibration curve contained eight CSs ranging from 2.2 to 600 μM of NAC (namely, 2.2, 6.3, 12.5, 18.8, 25, 150, 400 and 600 μM). Linearity was assessed by linear regression of the chromatographic peak area (mAU*min) as a function of NAC concentration (μM). The average back-calculated concentrations were also assessed. The slopes and Y-intercept of the curves where compared in order to assess reproducibility.

8.2.3.3 Lower and Higher Limits of Quantification

The LLOQ was defined as the lowest concentration of NAC that could be quantified with acceptable accuracy and precision. For this, six samples with a concentration of 2.2 μM were analyzed for the accuracy and the intra- and inter-assay precision. Additionally, the same was performed for the validation of the HLOQ.

8.2.3.4 Accuracy

Accuracy was defined as the closeness to the theoretical concentration of the QC samples. To study the accuracy of the method, six samples from the LLOQ and the HLOQ as well as four QC samples in between the concentration range (QC1 = 3.2 μM; QC2 = 9.4 μM; QC3 = 50 μM and QC4 = 300 μM). The accuracy of the method was calculated as the ratio between the measured and theoretical concentrations, expressed in percentage.

8.2.3.5 Intra- and Inter-assay Precision

The precision of the method was defined as the closeness between multiple measurements. Intra- and inter-assay precisions were assessed by analysis of six samples from LLOQ, HLOQ and QC. While intra-assay precision was assessed by the coefficient of variation (CV) obtained from analytical runs on the same day, the inter-assay precision was evaluated using the CV obtained on different days. Precision calculation was performed assuming that its value would ideally be 100%, subtracting the obtained CV.

8.2.3.6 Stability and Storage Conditions

Stability tests were carried out to ensure that the concentration of uNAC is not affected during sample preparation, sample analysis or even during storage. To this end, urine samples from volunteers were collected to a) readily quantify uNAC fractions; b) store at RT for 2 h or 6 h; c) store at 4 °C for 2 h or 6 h; and d) store at −80 °C and thaw. The stability was also investigated in samples stored at −80 °C for 1, 6 or 12 months after collection.

8.2.3.7 Quantification of uNAC vs. Free NAC

The ability to distinguish uNAC from free*N*-acetylcysteine in urine was also evaluated (Fig. 8.1). Thiols might be present in urine in different fractions, the protein-bound and the unbound/free fractions, which includes disulfides and free reduced forms [23]. To measure each fraction, we adapted the protocol previously described [20] (Fig. 8.1). The sum of all fractions corresponds to the total urinary NAC content, herein called uNAC. To quantify free reduced NAC that does not require the addition of a reduction step as it is present in its reduced sulfhydryl form, freshly collected urine samples from volunteers referred at Sect. 8.2.1 were firstly submitted to protein precipitation with TCA, followed by centrifugation (13,000 g, 10 min, at 4 °C). Then, samples were incubated with reverse osmosis water at RT for 30 min. The use of reverse osmosis water instead of the reduction agent TCEP allows the posterior derivatization of only the naturally free reduced NAC (Fig. 8.1). The protocol for the derivatization step was identical to the one described in Sect. 8.2.2.3. For quantification of total uNAC, a surrogate of mercapturates derived from CysSSX, samples from the same volunteers were analyzed by performing the protocol described in Sect. 8.2.2.3.

8.2.4 Method Applicability

8.2.4.1 uNAC in Urine of Rats and Mice

Experiments were performed using male Wistar Crl:WI (Han) (*Rattus norvegicus* L.) and male C57BL/6 mice obtained from the NOVA Medical School rodent facility. Animals were housed in polycarbonate cages with wire lids and maintained under standard laboratory conditions as follows: artificial 12 h light/dark cycles at RT. Animals were maintained ab libitum on a standard laboratory diet (SDS diets RM1 for rats and RM3 for mice, Special Diets Services) and reverse osmosis water. While the RM1 diet has 2.2 g/Kg of methionine and 2.4 g/Kg of cystine, the RM3 diet has 3.6 g/Kg of methionine and 3.5 g/Kg of cystine. Urine samples were collected throughout time. Animals were specific-pathogen-free (SPF) according to FELASA recommendations. All applicable institutional and governmental regulations concerning the ethical use of animals were followed, according to the NIH Principles of Laboratory Animal Care (NIH Publication 85 23, revised 1985), the European guidelines for the protection of animals used for scientific purposes (European Union Directive 2016/63/EU) and the Portuguese Law n° 113/2013. The experimental procedures received prior approval by the Institutional Ethics Committee of the NOVA Medical School for animal care and use in research (protocol n° 15/2017/CEFCM).

The levels of uNAC were obtained using the procedure described in Sects. 8.2.2.3 and 8.2.2.4.

8.2.4.2 uNAC in Patients with Kidney Disease

To obtain a proof-of-principle of the impact of kidney dysfunction on uNAC, its levels were quantified in urine samples from patients after an episode of acute kidney injury (AKI) through the same procedure described in Sects. 8.2.2.3 and 8.2.2.4. The levels of uNAC were then compared with those obtained in urine samples form volunteers (quantified in Sect. 8.3.3.7).

8.2.5 Statistical Analysis

Statistical analysis was performed using GraphPad Prism® version 5.0 (GraphPad Software Inc., San Diego, CA, USA). Data are presented in percentage (%), mean ± standard error of the mean (SEM) or median ± interquartile range (IQR), whenever applicable. Variability among data is expressed in CV. The *F*-test was used to explore differences between the slopes and elevations of the calibration curves and to ascertain the influence of the concentrations on the chromatographic signal area. Comparisons among groups were performed using paired t-test, unpaired t-test or one-way ANOVA, whenever applicable.

8.3 Results and Discussion

8.3.1 Method Development

To clarify uNAC as a surrogate of kidney tubular cell excretion of CysSSX by the action of NAT8, serum samples from five volunteers (median [IQR] age of 39 [28–54] years old, 60% men, all Caucasian) were analyzed using the protocol described in Sects. 8.2.2.3 and 8.2.3.7, respectively, for quantification of uNAC and free NAC. Both uNAC and NAC were not found in all serum samples tested (Fig. 8.2). This suggests that uNAC is a marker of *N*-acetylation of cysteine-disulfides in the kidney. It is known that cysteine-*S*-conjugates are acetylated by NAT8 that is predominantly expressed in human kidney tubular cells and present at substantially lower levels in hepatocytes [16]. The absence of uNAC in serum might indicate that the mercapturates from the liver are formed in small amounts and/or they are excreted into the bile and not found in circulation. As so, uNAC present in urine, i.e., the mercapturates from CysSSX (the oxidized fraction), is not derived from the glomerular filtrate but rather from tubular cells where the expression of NAT8 is the highest [16]. To the best of our knowledge, protein *N*-acetyl-cysteinylation by

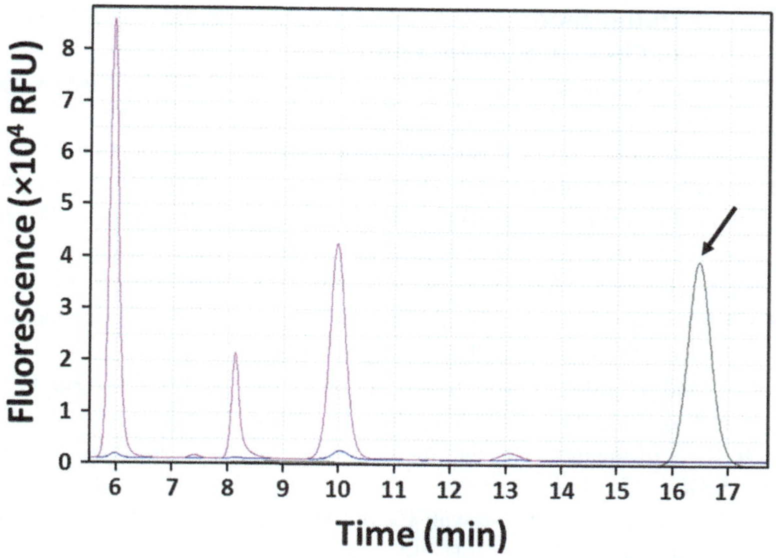

Fig. 8.2 Representative chromatograms of NAC in serum (black arrow). In black, a chromatogram for calibration solution; In pink, total fraction of uNAC (oxidized + reduced + protein-bound fractions) and in blue uNAC free reduced fraction in a serum sample of a volunteer

NAT8 has never been described in literature. However, as protein cysteinylation is described in kidney [2], these conjugates will be plausible substrates of NAT8. Nevertheless, as urine from healthy kidneys is not rich in proteins, we have not addressed this topic in this work, which should be evaluated in models of kidney disease.

Free reduced NAC was also not expected to be observed in accordance with previous reports showing that NAC levels are only detectable in serum or plasma samples after oral administration of the drug *N*-acetylcysteine [23].

In fact, the acetylation of Cys in the kidney seems to be the only process by which endogenous NAC is produced and, because of that, not present in plasma but only in urine. As so, the obtained results support that the approach herein employed enables to distinguish *N*-acetyl-CysSSX formed in tubular cells from the one formed in hepatocytes by measuring uNAC in urine.

8.3.2 Method Description

8.3.2.1 Optimization of Reduction and Derivatization Steps

The general approach for the quantification of thiols initially requires the generation of free thiols from disulfides, a process accomplished by the use of a reducing agent such as TCEP [18]. This step is required for quantification of the total and of the oxidized fractions of thiols (free and protein-bound, as this is the only way for the sulfhydryl groups to be available for posterior derivatization). Other authors have taken the advantage of the properties conferred by TCEP agent [24], but the quantification was performed in a reversed-phase ion-pair liquid chromatography with ultraviolet (UV) detection at 355 nm. Herein we minimize the interferences characteristic of absorption methods, especially in the UV range, that are more susceptible to interference from other biological substances than methods using FD as the one herein proposed [25, 26]. On the other hand, thiols have physicochemical properties including high polarity and water-solubility that makes their extraction from biological matrices almost impossible without a derivatization process [27]. As thiols are not spontaneously fluorescent [29], this step needs to be performed with a fluorescent probe that reacts selectively with the free sulfhydryl group of the thiol in order to generate a derivative that is strongly fluorescent, hence enabling a highly sensitive and precise measurement [19–21].

Taking the advantage of a previously reported method [20], we added NAC to the analytical run and we were able to reduce the initial volume of samples, CS and reagents to half of the initially used volume without any changes in the chromatographic peak area.

8.3.2.2 Optimization of Chromatographic Conditions

The development of this methodology required the optimization of several chromatographic conditions: a) mobile phase – isocratic elution using different concentrations of acetate buffer with MeOH in several proportions; b) stationary phase – two different length C18 columns were tested (LiChroCART 125 × 4 and C18 LiChroCART 250 × 4) with different temperatures (from 23 to 35 °C) and c) flow rate – a flow of 0.6 and 0.8 mL/min were tested. As far as we know, this is the first method developed with the objective of quantifying the mercapturates of CysSSX.

Using an isocratic elution with 1% of MeOH in 100 mM acetate buffer (pH 4.5) on a C18 LiChroCART 250 × 4 column at 29 °C, the chromatographic peak of NAC eluted at 16.5 min, totally separated from the retention time of the remaining thiols (Cys: 5.9 min, HCys: 8.1 min; CysGly: 10.0 min; CoA: 11.8 min and GSH: 13.1 min).

There are few reports that describe methods that allow the quantification of uNAC, even though laborious and requiring high amounts of samples and using other reducing and labelling agents [24, 28, 29]. However, it is important to highlight that these methods were not developed to investigate mercapturates of CysSSX nor their potential as non-invasive indicators of renal injury. Instead, the objective was to develop methods for pharmacokinetics studies of the drug *N*-acetylcysteine. One of the plausible reasons that might explain the absence of studies investigating the meaning of increased or decreased levels of mercapturates of CysSSX in urine, might be the lack of information regarding NAT8 and disulfide stress, that is recent as it is their relation with renal function [2, 14, 17].

The novelty of our approach was to associate disulfide stress and increased CysSSX to NAT8 activity and their detoxification as mercapturates. This is in fact the first report aiming at the quantification of uNAC as a surrogate of the urinary mercapturates derived from CysSSX and that provides evidence that its formation occurs in kidney.

8.3.3 Method Validation

8.3.3.1 Selectivity and Carry-over Effect

The injection of blank samples of PBS 1× or MeOH or mobile phase (100 mM acetate buffer (pH 4.5) and MeOH [99:1 (v/v)]) showed no interferences with uNAC. Additionally, the injection of three blank samples after the higher CS (600 μM) showed no carry-over effect.

8.3.3.2 Linearity

Linearity was assessed using three calibration curves containing eight CSs ranging from 2.1875 to 600 μM of NAC. Linear regression proved to be the most suitable model for fitting the experimental data (*Runs Test*, $p > 0.05$). The concentration of the standard samples significantly influenced the chromatographic signal area (*F* test $p < 0.001$). Additionally, no differences were found between the slopes and elevations of the calibration curves. The correlation coefficient, r^2, of the calibration curves were > 0.99, showing a good adjustment of all calibration curves. The 95% confidence interval for the intercept contained zero, indicating that the concentration of the CSs was the only factor with significant influence on the chromatographic peak area.

The observed back-calculated values were close to the expected ones at each tested concentration. The CVs were lower than 13% for higher and lower limits and 12% for the remaining standards.

8.3.3.3 Lower and Higher Limits of Quantification

The LLOQ was set at 2.1875 μM, and the CV of multiple measures was 5%. Although concentrations lower than 2.1875 μM were also tested, the chromatographic peaks were not distinguishable from background noise. This LLOQ allows the quantification of NAC at reported urinary levels [32]. The HLOQ was defined at 600 μM, with CV of several measures lower than 12%. The accuracy and intra- and inter-assay precisions of LLOQ and HLOQ are presented in Table 8.1.

8.3.3.4 Accuracy

The accuracy ranged from 87% to 114% for LLOQ and HLOQ and between 102% and 112% for QC samples (Table 8.1).

8.3.3.5 Intra- and Inter-assay Precision

The intra-assay precision was 96% and 97% for LLOQ and HLOQ, respectively, while for the remaining QCs it ranged from 95% to 98%. As for the results obtained for the inter-assay precision, the values were 88% and 95% for the LLOQ and HLOQ and between 95% and 98% for the QCs (Table 8.1). The obtained results further support the use of this method to quantify with accuracy and precision the levels of uNAC in biological samples.

8.3.3.6 Stability and Storage Conditions

The variations on the levels of total uNAC in urine sample subjected to different storage conditions are presented in Table 8.2. No differences were found between the concentrations obtained for all samples from the same volunteer for the several tested conditions. The conditions herein tested did not influence uNAC levels (One-way ANOVA).

Table 8.1 Mean accuracy, intra- and inter-assay precision of the method

Sample (μM)	Accuracy (%)	Precision (%)	
		Intra-assay	Inter-assay
LLOQ (2.2)	96	96	95
QC1 (3.1)	98	95	96
QC2 (9.4)	104	97	95
QC3 (50)	105	98	98
QC4 (300)	110	96	98
HLOQ (600)	102	97	88

LLOQ lower limit of quantification, *QC* quality control, *HLOQ* higher limit of quantification

The presented results are particularly important for clinical studies, where the logistics of sample procurement, its storage, lifecycle and deliver to the laboratory are most of the times difficult to manage one at a time. Similarly to mercapturates of endogenous CysSSX, the levels of several mercapturates derived from both endogenous (e.g. acrolein, 4-hydroxy-2-nonenal) and exogenous substances (e.g. flupirtine, 1-vinyl-2-pyrrolidones, thioTEPA, acrylamide, acrylonitrile, benzene, 1,3-butadiene, crotonaldehyde, ethylene, ethylene oxide, *N,N*-dimethylformamide, vinyl chloride, propylene oxide, styrene, toluene, propylene oxide) were found to be fairly stable when subjected to different storage conditions or several freeze-thaw cycles [33–39].

8.3.3.7 Quantification of uNAC vs. Free NAC

Urine levels of uNAC (58 [44–69] μM) and free NAC (9 [8–13] μM) from volunteers are presented in Table 8.3. The levels of free NAC represent 13–24% of total uNAC. (Paired t-test, $p = 0.001$).

The presence of endogenous free reduced NAC in urine samples was already reported in healthy volunteers [24, 28]. NAC levels in urine range from 3 to 7 μM, and represented 18–26% of total uNAC, in accordance with our findings. In fact, the acetylation of cysteine in the kidney seems to be the only process whereby endogenous NAC is present in urine [24, 28]. Although this represents a small contribution to uNAC, the presence of NAC in urine has been also described a result of oral administration of NAC [29].

Our results support the hypothesis that CysSSX might undergo acetylation by NAT8, forming mercapturates that are eliminated in urine. uNAC can thus be used as a surrogate of

Table 8.2 Stability of uNAC in urine samples from volunteers subjected to different storage conditions

	Storage conditions							
Temperature	RT		4 °C		−80 °C			
Time	2 h	6 h	2 h	6 h	1d	30d	180d	1y
uNAC[a] (%)	104 ± 5	107 ± 5	98 ± 4	113 ± 3	109 ± 4	117 ± 5	106 ± 6	97 ± 6

[a]Mean ± SEM; uNAC expressed as % of values quantified readily after collection; RT: room temperature

Table 8.3 Quantification of total uNAC vs. free reduced NAC in urine samples from volunteers

Sample	Age	Gender	uNAC (μM)	NAC (μM)	NAC (% of uNAC)
1	27	F	67	12	18
2	29	M	70	14	20
3	68	M	33	8	24
4	39	F	58	9	16
5	39	M	54	7	13

F female, *M* male, *uNAC* urinary surrogate of mercapturates from cysteine-disulfides, *NAC* free reduced *N*-acetylcysteine

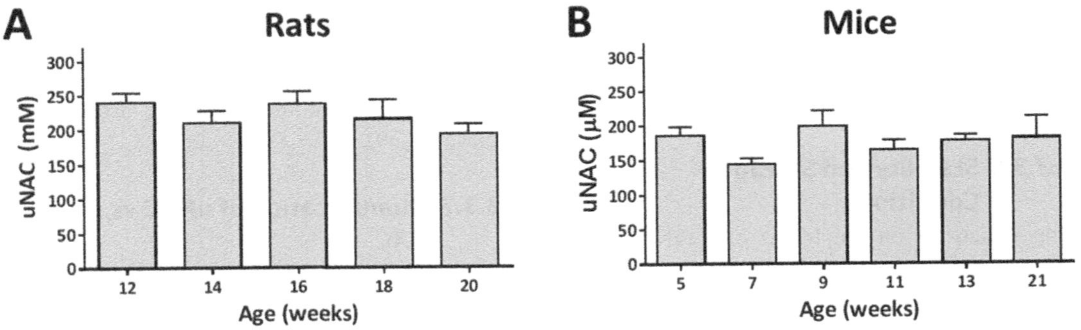

Fig. 8.3 Temporal variation of uNAC in urine samples from rats (**a**) and mice (**b**) (n = 5–13 per week)

these mercapturates and the differences between free reduced NAC and total NAC are derived from this acetylation step.

8.3.4 Method Applicability

8.3.4.1 uNAC in Urine of Rats and Mice

The levels of uNAC were quantified in urine samples from rat (n = 13 male rats, 12 weeks of age, mean ± SD body weight (BW) of 303 ± 28 g) and mouse (n = 13 male mice, 5 weeks of age, 19 ± 2 g of BW). There were no temporal differences in uNAC levels for both species (Fig. 8.3). The levels of uNAC found in both rat and mouse samples are higher than those quantified in human samples.

8.3.4.2 uNAC in Patients with Kidney Disease

uNAC was quantified in four patients after an episode of AKI. Anthropometric and clinical data are presented in Table 8.4. Lower levels of uNAC were found in urine of patients after an episode of AKI (23 [16–41] μM) in comparison with volunteers (58 [44–69] μM) (Unpaired t-test, p = 0.016) and all patients progressed to chronic kidney disease (CKD). These findings support that uNAC might represent an added value for the diagnosis and monitoring of kidney tubular injury.

Table 8.4 Anthropometric and clinical data from patients with an episode of AKI

Patient	A	B	C	D
Age (years)	74	85	56	23
Sex	M	F	F	M
Race	NB	B	NB	NB
Co-morbidities	DM, CVD	CVD	–	CVD
AKI etiology	CRSd	NSAIDs	RPGN (Wegener)	MHT, TMA
AKI SCr (mg/dL)	2.3	3	3.13	4.5
AKI eGFR[a] (mL/min/1.73m^2)	27	16	16	17
AKI Grade	2	1	3	3
Urine sampling time (days after AKI)	17	2	17	6
SCr (mg/dL)	2.95	2.19	3.12	4.31
eGFR (mL/min/1.73m^2)	20	23	16	18
uNAC(µM)	45	17	15	28

[a]calculated with CKD-EPI equation using serum creatinine; *AKI* acute kidney injury, *B* black, *CRSd* cardiorenal syndrome, *CVD* cardiovascular disease, *DM* diabetes mellitus, *eGFR* estimated glomerular filtration rate, *F* female, *M* male, *MHT* malignant hypertension, *NB* non-black, *NSAIDs* non-steroidal anti-inflammatory drugs, *RPGN* rapidly progressive glomerulonephritis, *SCr* serum creatinine, *TMA* thrombotic microangiopathy

8.4 Conclusions

The developed and validated HPLC-FD method is able to quantify the levels of mercapturates derived from endogenous CysSSX, herein called uNAC. This method might be applied in clinical studies aiming to investigate NAT8 and disulfide stress in kidney tubular dysfunction in man as well as in animal models of kidney tubular disease.

Acknowledgements This work was supported by Fundação para a Ciência e a Tecnologia: EXPL/DTP-FTO/1792/2013; iNOVA4Health - UID/Multi/04462/2013, a program financially supported by Fundação para a Ciência e Tecnologia / Ministério da Educação e Ciência, through national funds and co-funded by FEDER under the PT2020 Partnership Agreement is acknowledged (Ref: 201601-02-021). Authors supported by Fundação para a Ciência e a Tecnologia: PD/BD/105892/2014 for CGD; 022125 for JM; SFRH/BD/130911/2017 for MJC; PD/BD/114257/2016 for NRC; Programa Operacional Potencial Humano and the European Social Fund (IF/01091/2013) for AMMA.

References

1. Wang, W., Ballatori, N.: Endogenous glutathione conjugates: occurrence and biological functions. Pharmacol. Rev. **50**, 335–356 (1998)
2. Coelho, N.R., Gonçalves-Dias, C., Correia, M.J., Grácio, P., Serpa, J., Monteiro, E.C., et al.: Cysteine oxidative dynamics underlies hypertension and kidney dysfunction induced by chronic intermittent hypoxia. Adv. Exp. Med. Biol. **1071**, 83–88 (2018)
3. Abuyassin, B., Badran, M., Ayas, N.T., Laher, I.: Intermittent hypoxia causes histological kidney damage and increases growth factor expression in a mouse model of obstructive sleep apnea. PLoS One. **13**, e0192084 (2018)
4. Moreno, M.L., Escobar, J., Izquierdo-Álvarez, A., Gil, A., Pérez, S., Pereda, J., et al.: Disulfide stress: a novel type of oxidative stress in acute pancreatitis. Free Radic. Biol. Med. **70**, 265–277 (2014)
5. Grunwell, J.R., Gillespie, S.E., Ward, J.M., Fitzpatrick, A.M., Brown, L.A., Gauthier, T.W., Hebbar, K.B.: Comparison of glutathione, cysteine, and their redox potentials in the plasma of critically ill and healthy children. Front. Pediatr. **26**(3), 46 (2015)
6. Jones, D.P., Carlson, J.L., Mody, V.C., Cai, J., Lynn, M.J., Sternberg, P.: Redox state of glutathione in human plasma. Free Radic. Biol. Med. **28**, 625–635 (2000)

7. Szwergold, B.S.: α-Thiolamines such as cysteine and cysteamine act as effective transglycating agents due to formation of irreversible thiazolidine derivatives. Med. Hypotheses. **66**, 698–707 (2006)
8. Guan, X., Hoffman, B., Dwivedi, C., Matthees, D.P.: A simultaneous liquid chromatography/mass spectrometric assay of glutathione, cysteine, homocysteine and their disulfides in biological samples. J. Pharm. Biomed. Anal. **31**, 251–261 (2003)
9. States, B., Foreman, J.W., Segal, S.: Cysteine and glutathione levels in developing rat kidney and liver. Pediatr. Res. **22**, 605–608 (1987)
10. Gorgulho, R., Jacinto, R., Lopes, S.S., Pereira, S.A., Tranfield, E.M., Martins, G.G., et al.: Usefulness of zebrafish larvae to evaluate drug-induced functional and morphological renal tubular alterations. Arch. Toxicol. **92**, 411–423 (2018)
11. Deol, R., Josephy, P.D.: Acetylation of aromatic cysteine conjugates by recombinant human N-acetyltransferase 8. Xenobiotica. **47**, 202–207 (2017)
12. Veiga-da-Cunha, M., Tyteca, D., Stroobant, V., Courtoy, P.J., Opperdoes, F.R., Van Schaftingen, E.: Molecular identification of NAT8 as the enzyme that acetylates cysteine S-conjugates to mercapturic acids. J. Biol. Chem. **285**, 18888–18898 (2010)
13. Juhanson, P., Kepp, K., Org, E., Veldre, G., Kelgo, P., Rosenberg, M., et al.: N-acetyltransferase 8, a positional candidate for blood pressure and renal regulation: resequencing, association and in silico study. BMC Med. Genet. **9**, 25 (2008)
14. Gonçalves-Dias, C., Morello, J., Semedo, V., Correia, M.J., Coelho, N.R., Monteiro, E.C., et al.: The mercapturomic profile of health and non-communicable diseases. High Throughput. **8**, 10 (2019)
15. Grácio, P.C., Gonçalves-Dias, C., Lopes-Coelho, F., Monteiro, E.C., Serpa, J., Silva, C.L., Pereira, S.A.: Changes in N-acetyltransferase 8 in kidney tubular cell: Injury, recovery and mesenchymal stromal cell-based therapy. In: 6th IEEE Portuguese Meeting on Bioengineering, ENBENG 2019 – Proceedings [8692517]. Institute of Electrical and Electronics Engineers Inc. (2019)
16. Chambers, J.C., Zhang, W., Lord, G.M., van der Harst, P., Lawlor, D.A., Sehmi, J.S., et al.: Genetic loci influencing kidney function and chronic kidney disease. Nat. Genet. **42**, 373–375 (2010)
17. Gonçalves-Dias, C., Morello, J., Correia, M.J., Coelho, N.R., Antunes, A.M.M., Macedo, M.P., et al.: Mercapturate pathway in the tubulocentric perspective of diabetic kidney disease. Nephron. **143**, 17–23 (2019)
18. Burns, J.A., Butler, J.C., Moran, J., Whitesides, G.M.: Selective reduction of disulfides by tris(2-carboxyethyl)phosphine. J. Org. Chem. **56**, 2648–2650 (1991)
19. Imai, K., Toyo'oka, T., Watanabe, Y.: A novel fluorogenic reagent for thiols: Ammonium 7-fluorobenzo-2-oxa-1,3-diazole-4-sulfonate. Anal. Biochem. **128**, 471–473 (1983)
20. Grilo, N.M., Correia, M.J., Miranda, J.P., Cipriano, M., Serpa, J., Matilde Marques, M., et al.: Unmasking efavirenz neurotoxicity: time matters to the underlying mechanisms. Eur. J. Pharm. Sci. **105**, 47–54 (2017)
21. Nolin, T.D., McMenamin, M.E., Himmelfarb, J.: Simultaneous determination of total homocysteine, cysteine, cysteinylglycine, and glutathione in human plasma by high-performance liquid chromatography: application to studies of oxidative stress. J. Chromatogr. B Analyt. Technol. Biomed. Life Sci. **852**, 554–561 (2007)
22. European Medicines Agency: Guideline on bioanalytical method validation. EMA Guidel. **44**, 1–23 (2012)
23. Tsikas, D., Sandmann, J., Ikic, M., Fauler, J., Stichtenoth, D.O., Frölich, J.C.: Analysis of cysteine and N-acetylcysteine in human plasma by high- performance liquid chromatography at the basal state and after oral administration of N-acetylcysteine. J. Chromatogr. B Biomed. Appl. **708**, 55–60 (1998)
24. Hannestad, U., Sörbo, B.: Determination of 3-mercaptolactate, mercaptoacetate and N-acetylcysteine in urine by gas chromatography. Clin. Chim. Acta. **95**, 189–200 (1979)
25. Peng, H., Chen, W., Cheng, Y., Hakuna, L., Strongin, R., Wang, B.: Thiol reactive probes and chemosensors. Sensors (Basel). **12**, 15907–15946 (2012)
26. Kuśmierek, K., Chwatko, G., Głowacki, R., Bald, E.: Determination of endogenous thiols and thiol drugs in urine by HPLC with ultraviolet detection. J. Chromatogr. B Analyt. Technol. Biomed. Life Sci. **877**, 3300–3308 (2009)
27. Jakubowski, H., Głowacki, R.: Chemical biology of homocysteine thiolactone and related metabolites, in: G. Makowski (Ed.), Adv. Clin. Chem., Elsevier, 81–103 (2011)
28. Kuśmierek, K., Bald, E.: Determination of N-acetylcysteine and thioglycolic acid in human urine. Chromatographia. **67**, 23–29 (2008)
29. Ventura, P., Panini, R., Pasini, M.C., Scarpetta, G., Salvioli, G.: N-acetyl-cysteine reduces homocysteine plasma levels after single intravenous administration by increasing thiols urinary excretion. Pharmacol. Res. **40**, 345–350 (1999)

Diagnosis of Fluorosis by Analysis of Fluoride Content in Body Fluids Using Ion Selective Electrode Method

9

N. K. Mondal

Abstract

Analysis of fluoride in body fluids (urine and serum) is essential for fluorosis diagnosis. Although 24-h urine collection is adopted to assess community defluoridation/fluoride supplementation program/research studies, but not feasible for Clinical/Pathological laboratories. Patients are reluctant to bring 24-h urine samples. Hence, spot urine samples are collected in *clean, dry polypropylene bottles* (not glass) without any preservative and *analyzed* on the same day by *the Ion analyzer* (ISE method). *Equal volumes* of Total Ionic Strength Adjustment Buffer (TISAB) solution are then added with body fluids before analysis and mixed well to eliminate interference from other ions besides pH adjustment and to provide a constant ionic strength. Results are reported as mg of Fluoride/l (ppm). High fluoride level in body fluids is an indication of confirmed cases of fluorosis. Two *interventions, e.g.* withdrawal of fluoride intake and intake of nutritive diet was introduced for prevention and control of fluorosis. The present study is to provide useful guidelines for monitoring of fluorosis disease in human *beings, those* who are at the risk of fluoride exposure.

Keywords

Body fluids · Fluoride · Diagnosis · Fluorosis · Early detection

9.1 Introduction

Fluorosis disease is endemic in several countries of the *world, including* India [1–7]. In India, people are suffering from fluorosis due to ingestion of excess fluoride (F) through various sources [8–12] such as *the drinking* of untreated ground water [13], consumption of food products rich in F [14, 15], *the use* of fluoridated dental products and inhalation of dusts and fumes from industrial emission [16]. There *is* many F containing drugs that are often prescribed for long periods of time [17] viz. Sodium Fluoride (NaF) for Osteoporosis, Osteosclerosis, Dental caries besides anti-depressant, anti-inflammatory drugs, *anesthetic* agents, etc. Fluorosis can afflict *the multiple systems*, organs, tissues and cells of young and old without gender discrimination. There are several factors that are responsible for the onset of fluorosis in person *such* as duration of exposure, quantity of F consumed, age factor, nutritional and hormonal status of an individual. The people

N. K. Mondal (✉)
Fluorosis Foundation of India (Camping & Working for JICA Funded 'Fluorosis Mitigation Programme' in Nagaur District, Rajasthan, India),
New Delhi, India

H. M. Baptista Carreira dos Santos (ed.), *Translational Urinomics*, Advances in Experimental Medicine and Biology 1306, https://doi.org/10.1007/978-3-030-63908-2_9

living in rural or semi-urban areas of India are suffering from fluorosis due to ingestion of untreated ground water from *a tube* well, hand pump, bore well, etc. which contains F > 1.0 mg/l as a natural contaminant. As per Bureau of Indian Standards guideline (BIS) [18], the acceptable limit of F in drinking water is *up to* 1.0 mg/l, less the better and permissible limit is 1.5 mg/l in the absence of alternate source although WHO's permissible limits is 1.5 mg/l. Fluorosis is a complex public health problem [19]. Accumulation of high F in the human body causes derangement in metabolism and results in major health disorders like Dental [20, 21], Skeletal [22, 23] and Nonskeletal fluorosis [24–26] affecting teeth, bone and soft-tissues, respectively. Dental fluorosis is the earliest visible signs of chronic F toxicity, affects the children due to consumption of *the excess F* during developmental stages teeth up to 6–8 years. It causes irreversible *discoloration* or staining of the teeth with yellow/yellowish brown/deep brown horizontal streaks or spots which is bilaterally symmetrical and away from the gums. Skeletal fluorosis causes pain in major joints with rigidity in *the neck*, shoulder, back (*lumbar* spine), hip (pelvic region) and knee besides restricted movement, inability to squat. Severe form of Skeletal fluorosis results in marked disability, crippling and *paralysis*. Symptoms of Dental and Skeletal fluorosis [27] *appear* at late/advanced stages of F toxicity when there is very little scope for recovery or treatment. Nonskeletal fluorosis symptoms also known as 'Early Warning Signs' viz. *gastrointestinal* disorders like gas formation [28], pain in stomach [29], constipation, nausea, loss of appetite, *muscle* weakness, tiredness, headache, frequent tendency to urinate (polyuria) and excessive thrust (polydipsia) [30], *anemia* [31], allergic *reaction, etc.* are the earliest manifestations of *fluorosis* which affect the soft tissues and organs of the body. But the clinical symptoms of Nonskeletal fluorosis often overlap with other disorders in the body, hence misleads the Doctors to suspect and diagnose fluorosis. As a result, patients remain undiagnosed and continue to consume F unknowingly and finally lead to crippling and *paralyzed* conditions. Fluorosis causes irreversible and permanent damage to the body, therefore early detection through confirmed diagnosis is the most important factor [32, 33].

To prevent and control of fluorosis, it becomes necessary to develop a diagnostic procedure by which F toxicity manifestation could be detected *in theprevention* and control of fluorosis. Retrieval of clinical history, laboratory based diagnostic *tests, including* radiological investigations is the most important for early diagnosis of fluorosis. Fluoride contents of various body *parts, namely* urine, blood (serum), bones, nails, hair, saliva and breast milk can be used for analysis of F *content, but* most easily available body fluid i.e. urine and blood (serum) with minimum sample preparation is commonly preferred for analysis. Although there are several methods like Gas Chromatography, Atomic Absorption, Spectrophotometry, etc. are available for estimation of F, well-established Ion Selective Electrode (ISE) technology using potentiometric method is considered as 'Gold Standard Method' for its precision and accuracy [34, 35] and widely used for determination of F. The main aim of this communication is to provide information on how to diagnose fluorosis correctly by analysis of F content by following a rapid, simple suitable analytical method, suitable for clinical laboratories without much hassle [36–38].

9.2 Materials and Method

Retrieval of clinical history, laboratory based and radiological investigation *is* important for early diagnosis of fluorosis. A battery of diagnostic tests to be *conducted* for the *patients, those* who are having health complaints suggestive of F poisoning.

9.2.1 First Method: Estimation of SA/GAG Ratio

The first diagnostic procedure developed is based on estimation of bone matrix molecules viz. Glycosaminoglycan (GAG) and Sialic acid (SA) and assessment of *the SA/GAG ratio* in the blood for early detection/diagnosis of F toxicity [39]. It is established that *the ratio* of SA/GAG reduced by one-third of fluorosis affected *patients, but* no significant changes in other bone disorders. SA/GAG test, though excellent *but* time-consuming and requires highly skilled *personnel.*

9.2.2 Second Method: Estimation of F Levels in Body Fluids (Urine and Serum)

To overcome these lacunae, efforts continued to develop a simple procedure for diagnosis of fluorosis where report can be provided within a short span of time. The second diagnostic method based on estimation of F contents in urine and serum by using F Ion Selective Electrode (ISE) through *the potentiometric method* [40–45]. The procedure provides rapid, uncomplicated determinations and eliminates interference associated with other methods. In addition to *body* fluids' test (i) it is also required to test the F content of drinking water, if the source of drinking water of the patient is an untreated ground water (viz. *Hand* pump, tube well, bore well, etc.) and (ii) to take *forearm X-ray radiography* to view interosseous membrane calcification/wavy outline of the radius and ulna. ISE method is inexpensive, suitable for wide concentration measurement range and unaffected by sample *color* or turbidity and commonly use for measuring F content in numerous samples in *a clinical laboratory*. It is considered as 'Gold Standard Method' by the Medical fraternity/Researchers due to its selectivity to F ion and higher precision.

9.2.3 Details of ISE Method

9.2.3.1 Apparatus

Ion Meter/Ion Analyzer Fluoride content in body fluids *is analyzed* potentiometrically using F Ion Selective Electrode (ISE) attached to an Ion Meter or Ion Analyzer. Ion Selective Electrode (ISE) measures the concentration of free F ions (inorganic form) in the solutions. A combination type single electrode (an integrated sensing electrode and reference electrode together) *is generally preferred* to use to decrease the amount of required solution/reagents and to cut wastage as well.

Interference The analysis of ISE is affected by many analytical interference viz. Polyvalent cations and pH of the solution which could be *minimized* by adding a proper buffer.

Reagents Total Ionic Strength Adjustment Buffer (TISAB) is essential to add during analysis as it provides a constant background ionic strength, de-complexes F ions and adjusts the solution pH (5.0–5.5).

Sample Collection (a) Blood sample - Serum to be separated by centrifugation as soon as possible after drawing the blood and *analyzed* directly without further treatment except mixing of *the appropriate buffer solution*; (b) Urine sample - Although 24 h urine sample collection is ideal for F analysis, but not feasible for clinical/pathological laboratories. Sometimes patients are reluctant to bring 24-h urine sample. Hence, spot urine samples (fasting not required) are collected from patients for F testing.

Sample Preservation Analysis of the urine and serum samples are done as early as possible on the day of collection. If not possible, then it can be stored at 4 °C in the refrigerator.

Handling No special sample preparation is required for analysis [46]. During analysis of F, urine sample (3–5 ml) or serum sample (1–2 ml) mixed with equal volume of buffer solution and mixed thoroughly in a small polyethylene beaker (not in glass). TISAB II solution is added to standard solution/samples in a ratio of 1:1 and measured directly.

Calibration Since the ISE is capable to trace out *very* minute quantity (i.e. in ppm or mg/l level) of F from the solution, calibration of Ion meter is essential before starting any analysis. It is very important to check the electrode operation (slope). *The slope* is defined as the change in millivolts (m*V*) observed with every ten-fold change in concentration. Obtaining a slope value in the range of (−) 54 to (−) 60 m*V* during calibration is the best means to check the *electrode* operation. Calibration of ISE *is done* by using F standards that narrowly bracket the expected sample concentration. If the sample concentration is unknown, then calibrate Ion meter with *at least* three F standard solutions.

Analysis of Samples After calibrating the Ion meter, F content of the samples *is measured* directly by dipping the electrode into the sample [47, 48]. Calibration and all measurements to be carried out in ambient/room temperature to avoid erratic reading/results due to temperature fluctuation. Express F contents as mg of F per *liter* (mg/l) or parts per million (ppm).

Quality Control Quality control procedure to be followed to get accurate and reliable results during *an analysis*. It is advisable to run in between samples, known concentration of F standard solution for checking the accuracy of the calibration of the meter. The sensitivity of the electrode to be verified every 1–2 h, if there *are* any temperature fluctuations. Forearm X-ray radiography shall *take* to assess the presence of interosseous membrane calcification along the margin of the radius and ulna which *may vary* in shape from a wavy outline or thorny projections. The presence of calcification of the forearm *interosseous* membrane is useful for the differential diagnosis of Skeletal fluorosis from other bone disorders.

9.3 Results

After analysis of F content, the values are compared with normal reference range. If the samples showed high F contents in comparison to normal reference range, the disease is diagnosed as fluorosis. The normal reference range of F contents in urine and serum is considered to be 0.1–1.0 mg/l and 0.02–0.05 mg/l, respectively [49]. Higher circulatory and excretory levels of F suggest F *poisoning, which* leads to differential diagnosis of Nonskeletal fluorosis from other diseases with overlapping clinical manifestations. Suspect F consumption through water, *food, beverages*, use of habit-forming substances viz. Rock salt, black tea, use of fluoridated toothpaste or inhalation of F dust and fumes. Analysis of F content of drinking water provides a clue to identify the source of F entry into the *body, i.e.* whether from water, food or any other sources [50]. Pain in the joints viz. Neck, shoulder, back, hip and knee without visible sign of fluid accumulation could be due to F toxicity manifestations. The X-ray radiograph may reveal *an increase* in bone mass and bone density and calcification of ligaments, which is an indication of *the definitive diagnosis* of Skeletal fluorosis besides *the high level* of F in body fluids. Due to Dental*fluorosis, discoloration* on the enamel surface in the front row of the permanent teeth, rarely *on* milk teeth of the children can be seen as horizontal streaks or spots, away from the gums, bilaterally symmetrical which are visible to the naked eyes.

9.4 Discussion

It needs to diagnose the disease correctly through a well-defined protocol at the earliest onset of the disease. Once fluorosis is detected or diagnosed at *a very early stage*, it is possible to prevent the disease from further progressions through practicing interventions. *Fluorosis* can be prevented and controlled by introducing simple, easy to practice 2 dietary interventions [51, 52]. Intervention – 1 i.e. Diet Editing: To withdraw *the F entry* from all sources to arrest further progression of disease. This would be followed by Intervention – 2 i.e. Diet Counseling: To promote adequate intakes of essential nutrients, vitamins, antioxidants and micronutrients through dietary sources viz. Fruits, vegetables and dairy products; by improving the diet, the damage done by F to the body is repaired and maintained [53, 54].

Drugs/medicines containing essential nutrients and antioxidants are not recommended as supplementing the diet with nutrients provide better results and faster recovery than pharmaceutical products. Monitoring of the patients *is carried* out for impact assessment after practicing interventions. Withdrawal of the source(s) identified for F, re-testing of F levels is made after an interval of 4–6 weeks and compared with baseline value. The reduction in the level of F in body fluids shall result in the disappearance of health complaints emanating from fluorosis within a short span of time. Follow-up patients carried out till recovery until F level comes down *to* normal reference range. Patients should continue the nutrient rich dietary regimes to keep them healthy.

Institutionlization of diagnosis of fluorosis and *practice* of interventions should be promoted for the benefit of the communities living in endemic areas ail from diverse health complaints due to F poisoning.

****

Normal reference range of fluoride contents in body fluids and drinking water

Parameters	Normal reference range	Fluoride content which may lead to Fluorosis
F in urine	0.1–1.0 mg/l	> 1.0 mg/l
F in serum	0.02–0.05 mg/l	> 0.05 mg/l
F in drinking water[a]	Upto 1.0 mg/l	> 1.0 mg/l
Radiograph of forearm X-ray	–	Interosseous membrane calcifications & dense bone

[a]As per Bureau of Indian Standards (BIS)

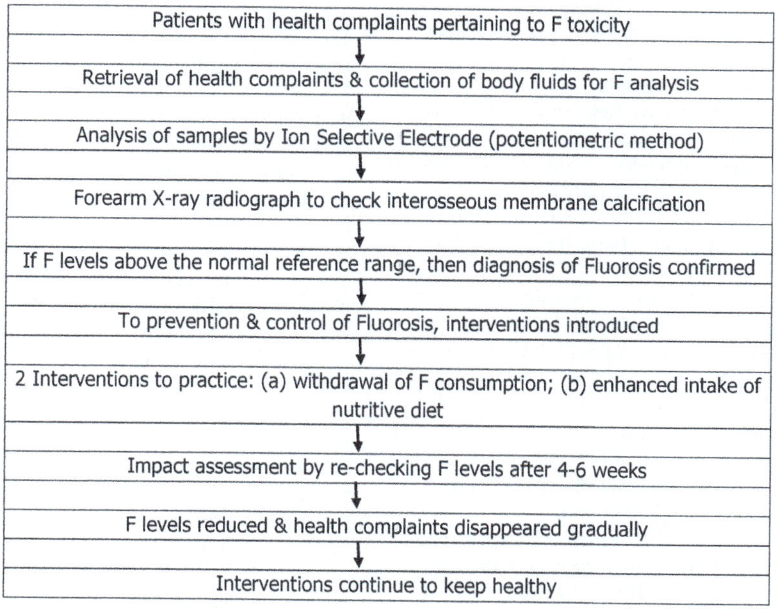

Schematic diagram of Fluorosis diagnosis protocol

References

1. Susheela, A.K.: A Treatise on Fluorosis (Revised 3rd Edition). Fluorosis Research & Rural Development Foundation, New Delhi (2007)
2. Susheela, A.K.: Fluorosis in developing countries: remedial measures and approaches. Proc. Indian Natn. Sci. Acad. (PINSA). **B68**(389–400) (2002)
3. Susheela, A.K.: Fluorosis and associated health issues. Indian J. Prac. Ped. **17**(2), 58–66 (2015)
4. Susheela, A.K., Sharma, K., Das, T.K.: Issues in fluoride toxicity and fluorosis. Ann. Natl. Acad. Sci. (India). **25**(1), 5–9 (1988)
5. Shortt, H.E., McRobert, G.R., Barnard, T.W., Mannadi, N.A.S.: Endemic fluorosis in the Madras presidency. Indian J. Med. Res. **25**, 553–568 (1937)
6. Arlappa, N., Qureshi, A.I., Srinivas, R.: Fluorosis in India – An overview. Int. J. Res. Dev. Health. **1**, 97–102 (2013)
7. Kanduti, D., Sterbenk, P., Artnik, B.: FLUORIDE – A review of use and effects on health. Mater. Sociomed. **28**(2), 133–137 (2016)
8. Buzalaf, M.A., Whitford, G.M.: Fluoride metabolism. Monogr. Oral Sci. **22**, 20–36 (2011)
9. Whitford, G.M.: The metabolism and toxicity of fluoride. Monogr. Oral Sci. **16**, 1–153 (1996)
10. Bond, A.M., Murray, M.M.: Kidney function and structure in chronic fluorosis. Br. J. Exp. Pathol. **33**(2), 168–176 (1952)
11. Villa, A., Anabalon, M., Zohouri, V., Maguire, A., Franco, A.M., Rugg-Gunn, A.: Relationships between fluoride intake, urinary fluoride excretion and fluoride retention in children and adults: an analysis of available data. Caries Res. **44**, 60–68 (2010)
12. Choubisa, S.L.: A brief and critical review of endemic hydrofluorosis in Rajasthan, India. Fluoride. **51**(1), 13–33 (2018)
13. Dharmaratne, R.W.: Fluoride in drinking water and diet: The causative factor of chronic kidney diseases in the North Central Province of Sri Lanka. Environ. Health Prev. Med. **20**, 237–242 (2015)
14. Nanda, R.S.: Fluoride content of North Indian foods. Indian J. Med. Res. **60**, 1470–1482 (1972)
15. Rao, K.V., Mahajan, C.L.: Fluoride content of some South Indian foods and their contribution to fluorosis. J. Sci. Food Agric. **51**, 275–279 (1990)
16. Susheela, A.K., Mondal, N.K., Singh, A.: Exposure to fluoride in smelter workers in a Primary Aluminium Industry in India. Intl. J Occup. Environ. Med. **4**, 61–72 (2013)
17. Baud, C.A., Boivin, G., Demeurisse, C.: Drug-induced skeletal fluorosis. Fluoride. **15**, 54–56 (1982)
18. Drinking water specification (IS: 10500). Bureau of Indian Standards (BIS), (2005)
19. Krishnamachari, K.A.: Skeletal fluorosis in humans: a review of recent progress in understanding of the disease. Prog. Food Nutr. Sci. **10**(3–4), 279–313 (1986)
20. Susheela, A.K., Bhatnagar, M., Gananasundaram, N., Saraswathy, T.R.: Structural aberrations in human Fluorosed teeth: biochemical and scanning Electron microscopic studies. Curr. Sci. **77**(10), 1677–1681 (1999)
21. Pandit, C.G., Radhavachari, T.N.S., Rao, D.S., Krishnamurti, V.: Endemic fluorosis in South India: a study of the factors involved in the production of mottled enamel in children and some bone manifestation in adults. Indian J. Med. Res. **28**, 533 (1940)
22. Susheela, A.K., Jha, M.: Fluoride ingestion and its influence on glycosaminoglycans in cancellous and cortical bone – A structural and biochemical study. Fluoride. **15**(4), 191–198 (1982)
23. Susheela, A.K., Jethanandani, P.: Serum haptoglobin and C-reactive protein in human skeletal fluorosis. Clin. Biochem. **27**(6), 463–468 (1994)
24. Dasarathy, S., Das, T.K., Gupta, I.P., Susheela, A.K., Tandon, R.K.: Gastroduodenal manifestations in patients with skeletal fluorosis. J. Gastroenterol. **31**, 333–337 (1996)
25. Kharb, P., Susheela, A.K.: Fluoride ingestion in excess and its effects on organic and certain inorganic constituents of soft tissues. Med. Sci. Res. **22**, 43–44 (1994)
26. Kaul, R.D., Susheela, A.K.: The muscle. In symposium on the non-skeletal manifestations of chronic fluoride toxicity. Fluoride. **9**(9) (1976)
27. Bruce, S.: The diagnosis of chronic fluoride intoxication including the use of serum and urinary fluoride ion levels and a forearm radiograph in the diagnosis of stage II and III skeletal fluorosis. Fluoride. **51**(1), 3–12 (2018)
28. Susheela, A.K., Das, T.K., Gupta, I.P., Tandon, R.K., Kacker, S.K., Ghosh, P., Deka, R.C.: Fluoride ingestion and its correlation with gastrointestinal discomfort. Fluoride. **25**(1), 5–22 (1992)
29. Das, T.K., Susheela, A.K., Gupta, I.P., Dasarathy, S., Tandon, R.K.: Toxic effects of chronic fluoride ingestion on the upper gastrointestinal tract. J. Clin. Gastroenterol. **18**(3), 194–199 (1994)
30. Das, T.K., Susheela, A.K.: Chronic fluoride toxicity and pituitary adrenal function. Environ. Sci. **1**, 57–62 (1991)
31. Susheela, A.K., Jain, S.K.: Fluoride toxicity: Erythrocyte Membrane Abnormality and "Echinocyte" formation. Fluoride Res. Stud. Environ. Sci. **27**(99), 231–239 (1986)
32. Susheela, A.K.: Fluorosis – Early warning signs and diagnostic test. Bull. Nutr. F. India. **10**(2) (1989)
33. Susheela, A.K., Mondal, N.K., Nalini, T., Gupta, R.: Early diagnosis and complete recovery from fluorosis through practice of interventions. J. Assoc. Physicians India. **62**, 572-579 (2014)
34. Martínez-Mier, E.A., Cury, J.A., Heilman, J.R., Katz, B.P., Levy, S.M., Li, Y., Maguire, A., et al.: Development of gold standard ion-selective electrode-based methods for fluoride analysis. Caries Res. **45**(1), 3–12 (2011)
35. Konieczka, P., Zygmunt, B., Namiesnik, L.: Comparison of fluoride ion-selective electrode based potentiometric methods of fluoride determination in

human urine. Bull. Environ. Contam. Toxicol. **64**, 794–803 (2000)

36. Basic methods for assessment of renal fluoride excretion in community preventive programmes for oral health. World Health Manual. ISBN 978 924 1548700, www.who.int (2014)
37. Mehta, A.: Biomarkers of fluoride exposure in human body. Indian J. Dent. **4**(4), 207–210 (2013)
38. Poydo-Szymaczek, P., Borysewicz-Lewicka, M.: Urinary fluoride levels for assessment of fluoride exposure of pregnant women in Poznan. Poland. Fluoride. **38**(4), 312–317 (2005)
39. Jha, M., Susheela, A.K., Krishna, N., Rajyalakshmi, K., Venkiah, K.: Excessive ingestion of fluoride and the significance of sialic acid: glycosaminoglycans in the serum of rabbit and human subjects. J. Toxicol. Clin. Toxicol. **19**(10), 1023–1030 (1982)
40. Yahyavi, H., Kaykhaii, M., Mirmoghaddam, M.: Recent developments in methods of analysis for fluoride determination. Crit. Rev. Anal. Chem. **46**(2), 106–121 (2016)
41. Hall, L.L., Smith, F.A., De Lopez, O.H., Gardner, D.E.: Direct potentiometric determination of total ionic fluoride in biological fluids. Clin. Chem. **18**(12), 1455–1458 (1972)
42. Tusl, J.: Direct determination of fluoride in human urine using fluoride electrode. J. Clin. Chem. Acta. **27**(1), 216–218 (1970)
43. Fuchs, C., Dorn, D., Fuchs, C.A., Henning, H., McIntosh, C., Scheler, F., Stennert, M.: Fluoride determination in plasma by ion selective electrodes – a simplified method for the clinical laboratory. Clinica. Chemica. Acta. **60**(2), 157–167 (1975)
44. Barnes, F.W., Runcie, J.: Potentiometric method for the determination of inorganic fluoride in biological material. J. Clin. Path. **21**, 668–676 (1968)
45. Kadioglu, Y., Gulaboglu, M., Ozturk, M., Dogan, N.: Determination of fluoride by using ion selective electrode (ISE) method in urine of healthy volunteer and patients before-after Sevoflurane Anesthesia. Lat. Am. J. Pharm. **34**(8), 1658–1663 (2015)
46. Milde, D., Novakova, K., Cermakova, I.: Fluoride determination in urine with ion selective electrode: within laboratory method and sample storage optimization. Acta. Univ. Palacki. Olomuc. Fac. Rer. Nat. Chemica. **43**, 104–109 (2004)
47. NIOSH Manual of Analytical Methods (NMAM), 4th Edition. FLUORIDE in urine: METHOD 8308. 2, 2–4 (1994)
48. Method 9214: Potentiometric determination of fluoride in aqueous samples with ion-selective electrode, part of test methods for evaluating solid waste, physical/chemical methods. https://www.epa.gov/sites/production/files/2015-12/documents/9214.pdf
49. Torra, M., Rodamilans, M., Corbella, J.: Serum and urine ionic fluoride: normal range in a nonexposed population. Biol. Trace Elem. Res. **63**(1), 67–71 (1998)
50. Standard Methods for the examination of water and wastewater, American Public Health Association (APHA), 21st Edition, (2005)
51. Mondal, N.K.: Diagnosis of fluorosis and recovery through easy to practise interventions. Fluoride. **51**(3), 230–242 (2018)
52. Susheela, A.K., Mondal, N.K., Tripathy, N., Gupta, R.: Early diagnosis and complete recovery from fluorosis through practice of interventions. J. Assoc. Phys. India (JAPI). **62**(7), 572–579 (2014)
53. Susheela, A.K., Bhatnagar, M.: Reversal of fluoride induced cell injury through elimination of fluoride and consumption of diet rich in essential nutrients and antioxidants. Mol. Cell. Biochem. **234/235**, 335–340 (2002)
54. Susheela, A.K., Rashmi, G., Mondal, N.K.: Anaemia in adolescent girls: An Intervention of diet editing and counselling. Natl Med. J. India. **29**(4), 200–204 (2016)

Dioctophimosis: A Parasitic Zoonosis of Public Health Importance

10

Soliane Carra Perera,
Carolina Silveira Mascarenhas,
Marlete Brum Cleff, Gertrud Müller, and
Josaine Cristina da Silva Rappeti

Abstract

Dioctophyme renale, the giant kidney worm, is a renal nematode from domestic and wild mammals that has zoonotic potential. In humans, dioctophimosis has been reported in several countries, mainly on the Asian continent, totaling more than 40 cases, which describe the parasite mainly infecting the kidneys, bladder, urethra and skin. Infection in animals and humans is related to the ingestion of the infective larva (L3) present in the aquatic oligochaete annelid (mandatory intermediate host) or fish and anurans (facultative paratenic hosts). Thus, the infection is related to the habit of drinking water contaminated with the mandatory intermediate host, as well as raw or undercooked meat from the facultative paratenic hosts. Dioctophimosis destroys the renal parenchyma and, in some cases, can cause the death of its hosts. In this chapter, we discuss the main topics regarding dioctophimosis in humans, domestic and wild animals, highlighting its importance in public health.

Keywords

Dioctophyme renale · Zoonosis · Human · Public health · Nematode

10.1 Introduction

Popularly known as giant kidney worm, *Dioctophyme renale* (Goeze, 1782) is a parasite nematode that infects domestic and wild mammals and, occasionally, humans [1, 2]. The parasitosis in animals is reported mainly in domestic dogs and in some species of wild carnivores such as ferrets, raccoons, maned wolfs, and Geoffroy's cat, among others [3–7]. Cases of dioctophimosis in humans have been reported in larger numbers in Asia [2]. The infection site of the parasite is the kidney of mammals, particularly the right-side one [1, 8]. Although uncommon cases have been seen of bilateral kidney parasitism by *D. renale* in both dogs and humans, which leads to worsening of the clinical setting and death [9, 10]. In animals, its occurrence has been reported in ecto-

S. Carra Perera (✉) · M. Brum Cleff
Group of Research, Teaching and Extension in Natural Products in the Veterinary Medical Clinic (FITOPEET), Department of Veterinary Clinics, Veterinary College, Federal University of Pelotas, Pelotas, RS, Brazil

C. Silveira Mascarenhas · G. Müller
Wild Animal Parasitology Laboratory (LAPASIL), Microbiology and Parasitology Department, Institute of Biology, Federal University of Pelotas, Pelotas, RS, Brazil

J. C. da Silva Rappeti
Project Dioctophyme renale in Dogs and Cats (PRODIC), Department of Veterinary Clinics, Veterinary College, Federal University of Pelotas, Pelotas, RS, Brazil

H. M. Baptista Carreira dos Santos (ed.), *Translational Urinomics*, Advances in Experimental Medicine and Biology 1306, https://doi.org/10.1007/978-3-030-63908-2_10

pic sites such as the abdominal cavity, thoracic cavity, urethra, bladder, uterus, ovary, mammary gland, scrotum, testicle, stomach, liver, subcutaneous tissue, and spine [1, 10–20]. The parasite has been found in humans in the skin, bladder, urethra, and kidneys, mainly causing lumbar pain and hematuria [2, 21].

Dioctophyme renale has blood-red color, sexual dimorphism, and may reach up to one meter long in its adult form (Fig. 10.1). The cycle of this helminth involves the participation of aquatic organisms such as oligochaetes (mandatory intermediate hosts), fish and anurans (facultative paratenic hosts). The definitive hosts (domestic and wild mammals) eliminate worm unembryonated along with urine and contaminate aquatic environments, where the first-stage larva develops (L1). Eggs with L1 are ingested by the intermediate host, where the infectant larva develops (L3). Fish and anurans may be infected by ingesting oligochaetes containing the L3, thus becoming paratenic hosts of infectant larvae, which remain mainly encapsulated in the serosa of stomach and intestine, and in the mesentery. The definitive hosts are infected by ingesting L3 larvae present in oligochaetes, fish, or anurans. The parasite then reaches the kidneys (preferential infection site), where it completes the cycle by laying eggs that will be eliminated by urine (Fig. 10.2). Humans are considered accidental hosts and can be infected in the same way as animals [1, 13, 22].

Dioctophyme renale is a cosmopolitan nematode, i.e., it has been reported infecting animals and humans in several countries and continents, such as North and South America, Europe, Africa, Asia, and Oceania [2, 24–29] (Fig. 10.3). Brazil is the country with the highest incidence of

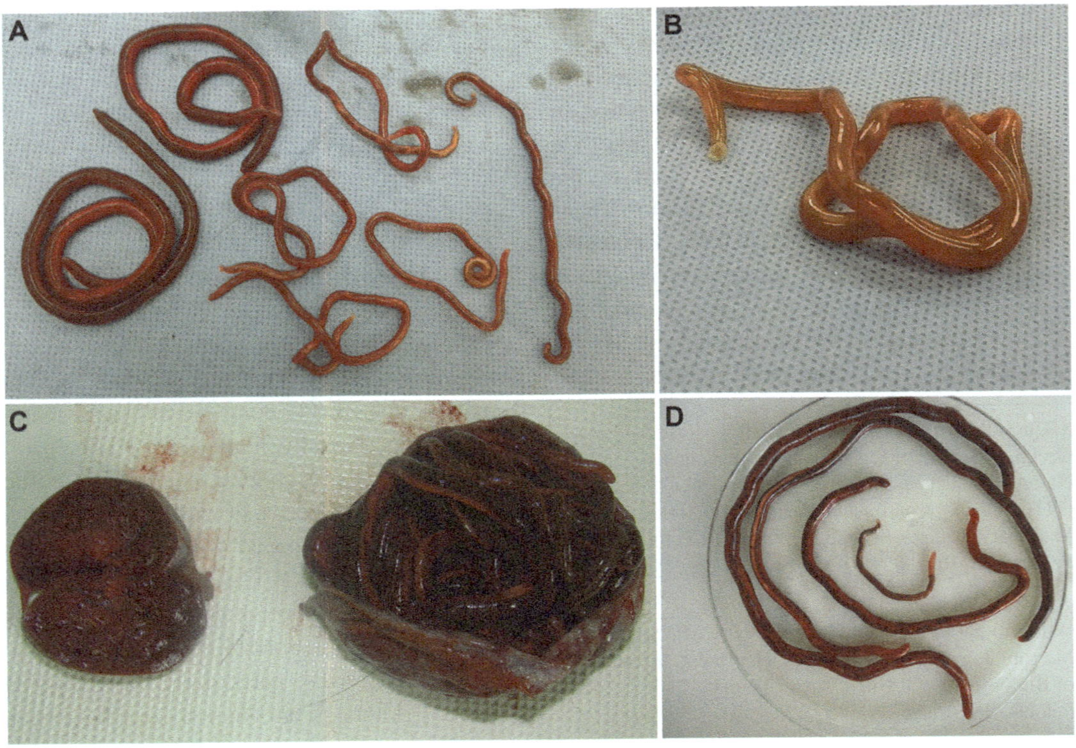

Fig. 10.1 (**a**) Two females (left) and five males (right) of *Dioctophyme renale* removed from the right-side kidney and abdominal cavity of dogs with dioctophimosis. (**b**) Male specimen of *D. renale*. Source: Project *Dioctophyme renale* in Dogs and Cats (PRODIC), Veterinary Clinics Hospital, Federal University of Pelotas, Rio Grande do Sul, Brazil. (**c**) Left- and right-side kidneys of lesser grison (*Galictis cuja*) infected by *D. renale*. Right-side kidney full of worms inside. Source: Albano, A.P.N. (**d**) Three females and one male of *D. renale* removed from the right-side kidney of lesser grison (*Galictis cuja*). Source: Mascarenhas, C.S

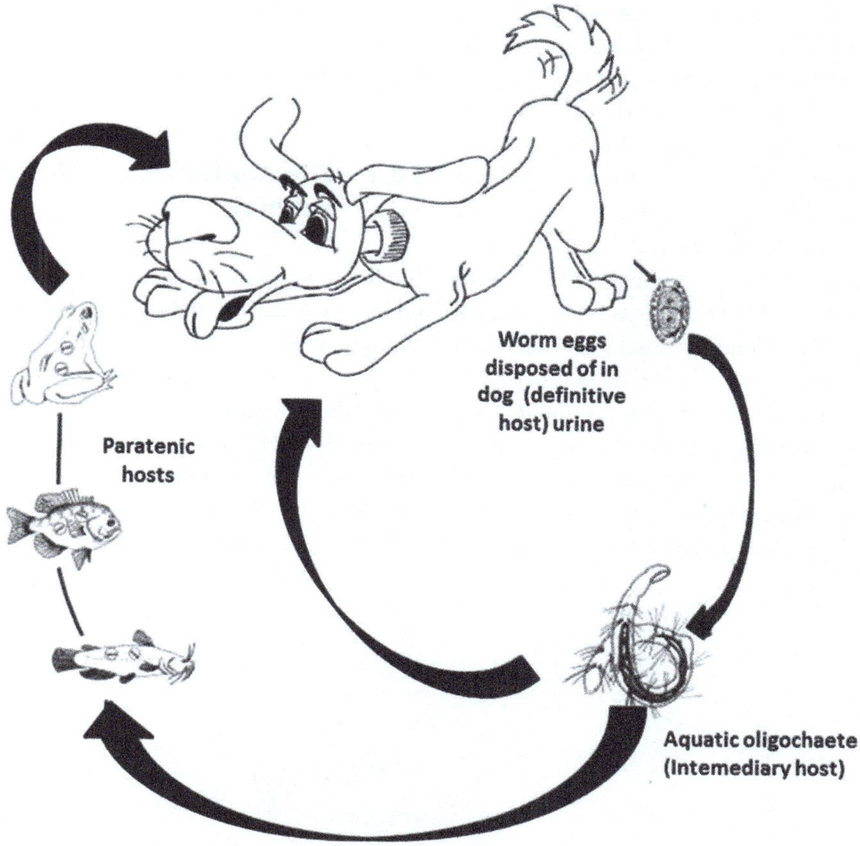

Fig. 10.2 Lyfe cycle of *Dioctophyme renale*. Source: Adapted from Anderson [23]

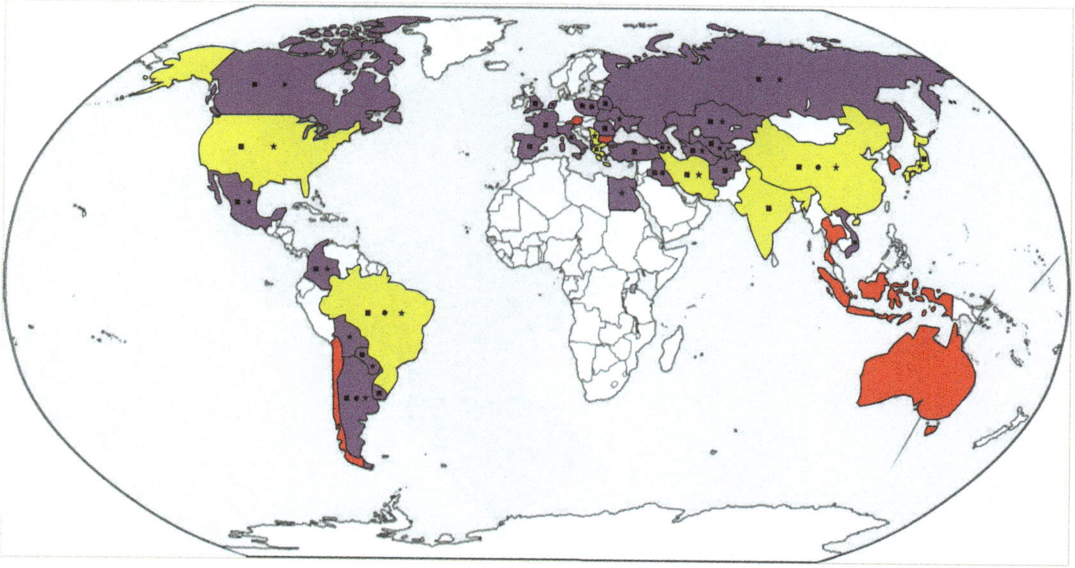

Fig. 10.3 Current geographic distribution of dioctophimosis in humans and animals. Red: *D. renale* in humans; Purple: *D. renale* in animals; Yellow: *D. renale* in humans and animals; ■: Dogs; ●: Cats; ★ : Wild animals

dioctophimosis in domestic and wild animals, which is found in over half of the states, i.e., Maranhão, Pará, Pernambuco, Minas Gerais, Espírito Santo, Mato Grosso, Mato Grosso do Sul, Goiás, São Paulo, Rio de Janeiro, Paraná, Santa Catarina, Rio Grande do Sul, and the Federal District [3, 4, 12, 17, 30–49] (Fig. 10.4). The state of Rio Grande do Sul stands out as it has the highest rate of animal infection, with diagnosed cases in Porto Alegre, Uruguaiana, Santa Maria, Passo Fundo, Capão do Leão, and Pelotas [7, 19, 48–54]. In cities in southern Rio Grande do Sul state, such as Pelotas, the parasitosis has been reported mainly in domestic dogs [6, 15, 51] and cats [41, 51]. In addition, wild animals have been reported as hosts of adult worms [4, 7, 52] and larval forms of *D. renale* [55–59]. Such occurrences as a whole are worrisome for the public health status of the region due to the zoonotic potential of the nematode.

10.2 Dioctophimosis in Domestic Animals

Studies published to date, in addition to those underway (data not published), indicate that the southern region of Rio Grande do Sul, Brazil, is the area with the most reported cases of dioctophimosis in domestic animals worldwide, mainly in dogs [6, 51]. Cases of canine dioctophimosis have also been reported in neighboring some countries of Brazil, i.e., Uruguay, Paraguay, Colombia, and Argentina [4, 18, 60–62]. The

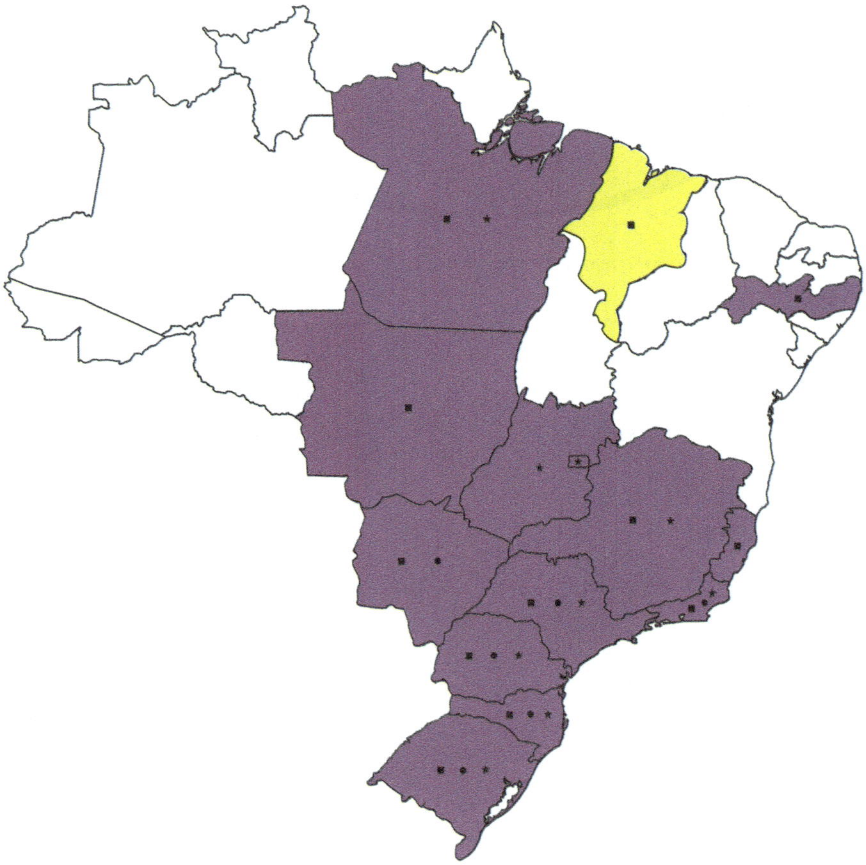

Fig. 10.4 Current geographic distribution of dioctophimosis in humans and domestic and wild animals in the states of Brazil. Purple: *D. renale* in animals; Yellow: *D. renale* in humans and animals; ■: Dogs; ●: Cats; ★: Wild animals

occurrence of infections in those countries indicates their climate and geographic characteristics may favor the maintenance of the parasitosis. Many cases occur in animals living in wetlands or close to large bodies of waters, such as rivers, creeks, and ponds [33, 61]. In addition to South America, the parasitosis in dogs has also been reported in North America, Europe and Asia [29, 63, 64, 65, 66, 67, 68, 69] (Fig. 10.3).

Dogs are the animals with the highest number of cases of dioctophimosis worldwide [19, 33, 51, 61, 70, 71] since access to the street observed in diagnosed patients and the little-selective feeding habits of the species may favor the ingestion of paratenic hosts and water contaminated with the intermediate host (oligochaete) [48]. Although the parasitosis is less common among domestic cats, its occurrence has been reported in Brazil, Argentina, Poland, China, and Iraq [29, 32, 38, 72, 73]. In Brazil, such infections have been diagnosed in the states of Mato Grosso do Sul, Rio de Janeiro, São Paulo, Paraná, Santa Catarina, and Rio Grande do Sul [6, 29, 32, 38, 41, 51]. According to Rappeti et al. (2017), 92 cases in dogs and three in cats have been diagnosed between 1981 and 2015 in Pelotas (Rio Grande do Sul), 81 of which since 2010. Perera et al. (2017a) analyzed 42 urine samples from dogs and one from a cat from Pelotas and found eight positive samples for *D. renale*, including the feline sample.

Infections by *D. renale* in domestic animals often result in the destruction of the renal parenchyma with atrophy and loss of glomeruli and renal tubules, fibrosis, and thickening of the renal capsule, which may contain parasites, serosanguineous fluid, and helminth eggs [10, 14, 74]. Several authors have reported progressive destruction of the cortical and medullar kidney layers, which reduces the organ to a fibrous capsule [73, 75] (Fig. 10.5a). The parasitosis may also cause hydronephrosis caused by flattening of the kidney crest and compression of the medullar and cortical regions by adult nematodes [19]. When present in the abdominal cavity, *D. renale* may cause peritonitis, alteration in omentum color, and thickening of the peritoneum with micronodullar aspect [12] (Fig. 10.5b).

Clinical signs vary according to parasite location. The animals are often asymptomatic, especially when neither or only one kidney is infected due to the compensation by the contralateral organ [43, 76]. However, when present, clinical signs manifest as hematuria and kidney pain due to the destruction of the parenchyma in the organ [19, 73]. Occasionally, some animals exhibit generalized discomfort, such as arching the back, abdominal pain, reluctance to walk, polyuria, strangury, vomit, and progressive weight loss, among others [8, 31, 77].

The surgical removal of helminths in infected animals has been recommended since the parasites may reach up to one meter in length [8].

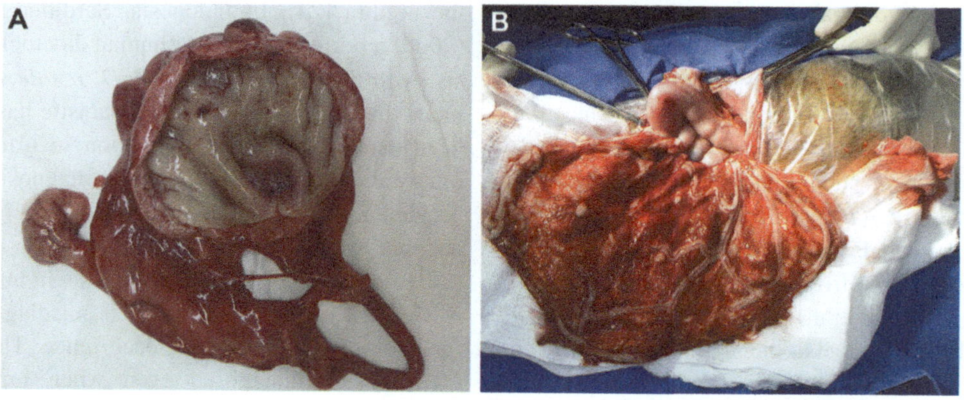

Fig. 10.5 Lesions caused by *Dioctophyme renale* in dogs. (**a**) Right-side kidney with destruction of the renal parenchyma. (**b**) Thickened omentum with micronodullar aspect. Source: Project *Dioctophyme renale* in Dogs and Cats (PRODIC), Veterinary Clinics Hospital, Federal University of Pelotas, Rio Grande do Sul, Brazil

Thus, when present in the kidney, nephrectomy is the treatment of choice when complete parenchyma destruction occurs [74]. When parasites are found in both kidneys, nephrotomy is the recommended procedure [10].

10.3 Dioctophimosis in Wild Animals

In addition to Brazil, dioctophimosis in wild animals has been reported in several countries such as Paraguay [78], Mexico [79], Canada [80, 81], United States [82–86], Egypt [28], Iran [88], and Japan [89], among others [29] (Fig. 10.3). The records refer to the helminthiasis in severeal species, including American marten (*Martes americana*), long-tailed weasel (*Mustela frenata*), mink (*Mustela vison*), American mink (*Neovison vison*), coyote (*Canis latrans*), gray fox (*Urocyon cinereoargenteus*), red fox (*Vulpes vulpes*), maned wolf (*Chrysocyon brachyurus*), raccoon (*Procyon lotor*), brown rat (*Rattus norvegicus*), mice (BALB-mice), and harbor seal (*Phoca vitulina*).

In Brazil, adult worms have been reported in wild mammals such as maned wolf (*Chrysocyon brachyurus*) [46], lesser grison (*Galictis cuja*) [4, 39, 40, 90], ring-tailed coati (*Nasua nasua*) [3], capuchin monkey (*Cebus apella*) [45], crab-eating fox (*Cerdocyon thous*) [37], neotropical otter (*Lontra longicaudis*) [52], and Geoffroy's cat (*Leopardus geoffroyi*) [7], among others [29]. Some wild animals have also been reported as possible paratenic hosts of third-stage *D. renale* larva, particularly yellow cururu toad (*Chaunus ictericus*) in Três Barras (Santa Catarina) [91], tamuatá catfish (*Hoplosternum littorale*), D'Orbigny's slider and Saint-Hilaire's side-necked turtle (*Trachemys dorbigni* and *Phrynops hilarii*), and Patagonia green racer (*Philodryas patagoniensis*) from Pelotas and Capão do Leão (Rio Grande do Sul) [56–59].

The occurrence of dioctophimosis in wild animals is a health public concern since mammals disseminate the eggs that contaminate aquatic environments, the dwelling of oligochaetes, fish, and turtles, which may host the infectant form of the parasite. The source of infection in humans is possibly the same as in animals, i.e., the ingestion of water contaminated with infected aquatic oligochaetes and the consumption of raw fish [2, 59]. Mascarenhas et al. (2019) raised a warning for the lack of studies assessing the presence of infectant *D. renale* larvae in fish, highlighting the importance of such investigation particularly in areas where fishing is practiced and the incidence of the parasitosis in animals is high.

10.4 Dioctophimosis in Humans

Dioctophyme renale is an important helminth for public health due to its zoonotic potential. Cases of human infection by it in the tegument, urinary vesicle, ureter, and kidneys have been reported likely due to the consumption of frogs, raw or rare fish, and non-boiled water [2, 9, 21, 92–96]. Most reports of human dioctophimosis come from Asian countries, such as China, India, Japan, Iran, Thailand, Indonesia, and South Korea, but other continents have also documented the parasitosis in humans, such as Europe (Greece, Yugoslavia, Austria, and Bulgaria), Oceania (Australia), North America (United States), and South America (Brazil and Chile), with a total of more than 40 cases [2, 24, 29, 93, 94] (Fig. 10.3).

In Brazil, a single case was reported in 1945, in which a woman eliminated the worm through the urethra [24]. In Indonesia, Sardjono et al. (2008) reported one case of human dioctophimosis, with the elimination of 23 *D. renale* specimens through the urethra. The parasite has also been detected through elimination in urine by other authors [9, 97]. In both cases, diagnosis was accidental since the clinical suspicion in the patients was not dioctophimosis. Yang et al. (2019) suggested that dioctophimosis in humans is not well recognized yet and is misdiagnosed for being considered a rare occurrence. Thus, it often goes undiagnosed even when imaging exams, such as ultrasound, are performed [98].

The most common infection sites in humans are the kidneys, mainly the right-side one [2, 9, 96]. Clinical signs and symptoms in humans, as well as in animals, are non-specific and are usually related to renal alterations such as hematuria, pyuria, lumbar pain, urinary retention, and fever [9, 99–101]. Possibly due to its rarity in humans, there is no standard treatment for dioctophimosis. No assessment has been made of the use of anthelmintics [94], although they are used in some cases, as is nephrectomy [21, 102]. Moreover, in extreme cases of renal lesions, the parasitosis may lead to death [9]. Prophylaxis and control measures recommended for humans, as well as for animals, are not drinking untreated water, especially from canals and ditches, thus preventing the ingestion of infected oligochaetes, and avoiding the consumption of raw or rare fish and frogs, which may host infectant larvae of the helminth [56, 59, 91, 103].

10.5 Diagnostic

Dioctophimosis may be diagnosed through urine tests, imaging exams such as abdominal ultrasound and computed tomography, and surgical and necropsy findings [6, 18, 19, 104]. Urinalysis is one of the most commonly employed methods to diagnose dioctophimosis as it allows identifying *D. renale* eggs in urinary sediment [8]. It is important the urine sample be homogenized prior to the exam since the eggs tend to settle at the bottom of collection flasks. Failure in doing so increases the risk of eggs going unnoticed in the sample.

After homogenization, centrifugation of 10 to 15 mL of urine at 1500 rpm is performed for five to 10 min [105]. In case the analysis aims to only detect *D. renale* eggs, centrifugation may be performed at 3000 rpm for 5 min. After centrifugation, the supernatant must be discarded and an aliquot of the urinary sediment placed onto a slide and analyzed in an optical microscope (10X and 40X) to visualize egg morphology [106].

Compared to eggs of other nematodes, *D. renale* eggs are relatively large at around 73–83 μm by 45–47 μm [8] and easily differentiated due to their morphological characteristics, i.e., elliptical shape; brown, yellow, or clear color; thick and wrinkled skin; and the presence of light-colored symmetrical bipolar ends [6, 107]. In addition, the core of the eggs is usually composed of only one or two cells when eliminated in urine [107] (Fig. 10.6).

Mature *D. renale* females usually lay eggs in lumps daily [108], which allows easily identifying eggs in urine. However, they may not be identified in urinalysis when the females are not laying eggs. That may occur when the parasite is unviable or dead inside the kidney or if she is too young and sexually immature, which prevents the production and dissemination of eggs.

Urine may be collected through spontaneous urination, bladder catheterization, or cystotomy [109]. Collection through spontaneous urination is non-invasive [105], but sufficient volume may be difficult to achieve in veterinary medicine since animals often halt urination at the moment of collection due to the proximity of the collector, little production depending on the time of collection, or lack of practice by the collector. Bladder catheterization is usually effective, but more traumatic [105] and is more easily performed in male dogs. Cystotomy may be carried out for urine analysis, in which case the bladder must be full and palpable, and collection may be guided by ultrasound [105]. Both bladder catheterization and cystotomy reduce the risk of microbial contamination of the sample, but require more skill [110].

The urine samples collected for urinalysis and *D. renale* egg identification must be stored under refrigeration and analyzed within one to two hours after collection due to microbial growth [110, 111]. However, if the samples are used only for identification of eggs, the analyses may be carried out a few days after collection. In that case, preservatives such as formalin may be used to reduce microorganism growth (bacteria and fungi) in the samples [110] with no impact on helminth eggs. Another major advantage of this diagnosis method is that only three parasite forms can be identified in the urine, i.e., *Dirofilaria immitis* microfilaria and *Capillaria plica* and *D. renale* eggs [109]. The latter two parasites are

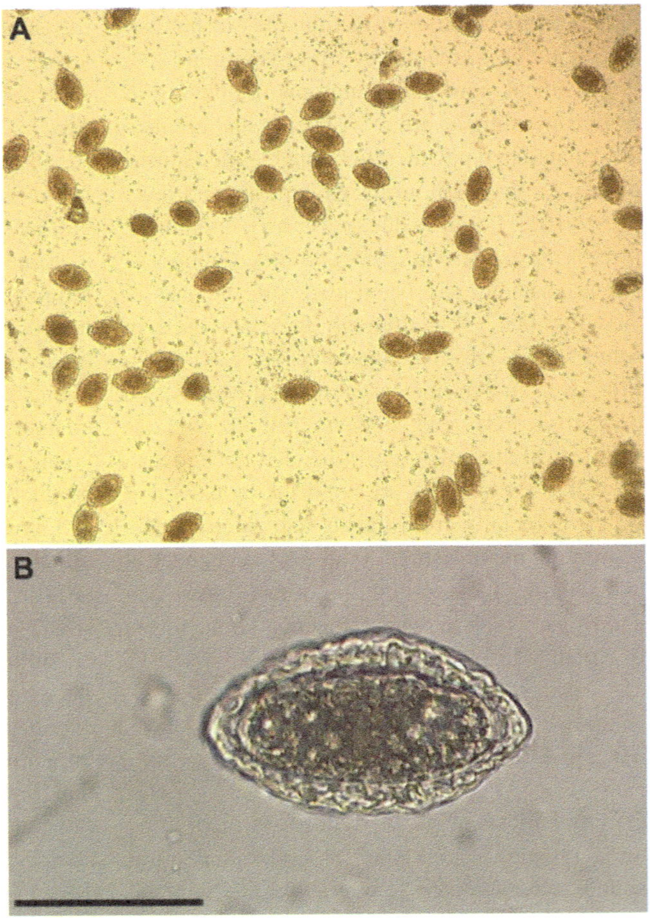

Fig. 10.6 (**a**) *Dioctophyme renale* eggs identified in the urinary sediment of a dog with dioctophimosis in an optical microscope at 10X. (**b**) Unembryonated *D. renale* egg (bar = 45 μm). Source: The author

able to infect the urinary system of animals, with *D. renale* infecting the kidney and *C. plica*, the bladder [112]. Furthermore, the helminths are morphologically different while their eggs are similar, except mainly for their sizes. *Capillaria plica* eggs are colorless and oval (barrel shaped) with bipolar ends, being 50–68 μm in length and 22–32 μm in width [112, 113], smaller than *D. renale* eggs.

The main advantages of diagnosing dioctophimosis through urinalysis is that it is an economic test that is quickly and easily performed by technicians trained to recognize the parasitosis. Thus, the experience of the professionals is key to differentiate *D. renale* eggs from those of other helminths to avoid false-positive and false-negative diagnostics. Urinalysis has been considered the golden standard to diagnose dioctophimosis [8]. However, despite being quick and convenient, it requires the presence of at least one viable sexually mature *D. renale* female infecting the kidney of the definitive host [6, 19]. It is, therefore, noteworthy that urinalysis is not a good diagnosis method in cases in which only ectopic parasitism occurs or when kidneys are infected with only male or unviable or sexually immature *D. renale* females.

Hence, in order to confirm diagnostic and helminth location, it is recommended to associate urinalysis with an imaging exams, of which ultrasound is the most commonly employed in veterinary medicine [51, 104]. Worms can be detected in such exam from their morphology given their sizes, regardless of sex, sexual maturity, or viability. In ultrasound images, helminths show a hypoechoic core and hyperechogenic edges [19]. In addition, the exam detects helminths both in the kidney and in ectopic sites such as the abdominal cavity, which is also often infected in animals [12, 19]. Nonetheless, false-positive and false-negative diagnostics may occur since the parasites could be mistaken for similar structures such as the intestine [104], or not be visualized if they are too young and small as they may overlap with anatomic structures. Besides echography exams, computed tomography has been used to diagnose renal dioctophimosis in animals [104]. Although effective, that is an uncommon exam in veterinary medicine in cases of *D. renale* parasitism and it is more often performed in humans, as are magnetic resonance imaging exams [2, 21, 27, 102].

Especially in animals, diagnosis is often carried out through surgical findings and necropsy [19, 48, 51]. That tends to occur when the hosts are asymptomatic, particularly when the worms are found in the abdominal cavity, or when the animal dies before being diagnosed with dioctophimosis [19, 48]. In humans, the helminthiasis is often diagnosed after nematodes or their fragments are eliminated in urine [9, 100].

In addition to urinalysis, imaging exams, and surgical and necropsy findings, Brazilian and Argentinian researchers have developed molecular and serologic assays to diagnose dogs with dioctophimosis for the presence of *D. renale* antibodies in the organism. Although serologic diagnosis is not routine in clinical practice, studies involving said method have shown promising results, as have molecular discoveries related to pseudocoelomic *D. renale* proteins [114–116].

10.6 Conclusion

Dioctophimosis is a parasitosis that may cause severe and irreversible lesions especially to the kidney of animals and humans. Moreover, as it is often a silent infirmity of chronic evolution and with unspecific clinical signs, the parasitosis may be improperly diagnosed or diagnostic may be late, leading to greater impact to the definitive hosts or, in some cases, even death. Therefore, it is highly important that healthcare professionals know the parasitosis so as to suspect it in differential diagnosis in animals and persons living in areas where infection occurs. Urinary sediment and imaging exams, particularly ultrasound, must be associated to diagnose the parasitosis and detect helminths in the kidneys and other infection sites.

After dioctophimosis is confirmed, proper treatment must be sought as soon as possible in order to minimize the damages caused to patients. It is also important to train human and animal healthcare providers regarding the control and prevention of *D. renale* given its zoonotic potential, in particular in regions where the helminthiasis is common.

Acknowledgments This study was financed in part by the Coordenação de Aperfeiçoamento de Pessoal de Nível Superior - Brasil (CAPES) – Finance Code 001. Therefore, we thank CAPES for financial support. We also thank the Program of Veterinary Medicine (PPGV) and the Program of Microbiology and Parasitology (PPGMPar) at the Federal University of Pelotas (UFPel), as well as the Project *Dioctophyme renale* in Dogs and Cats (PRODIC) and Dr. Ana Paula Neuschrank Albano for the availability of the images.

References

1. Measures, L.N.: Dioctophymatosis. In: Samuel, W.M., Pybus, M.J., Kocan, A.A. (eds.) Parasitic Diseases of Wild Mammals, 2nd edn, pp. 357–364. Iowa State University Press, Iowa (2001)
2. Yang, F., Zhang, W., Gong, B., et al.: A human case of Dioctophyma renale (giant kidney worm) accompanied by renal cancer and a retrospective study of dioctophymiasis. Parasite. **26**, 1–8 (2019). https://doi.org/10.1051/parasite/2019023
3. Milanelo, L., Moreira, M.B., Fitorra, L.S., et al.: Occurrence of parasitism by Dioctophyma renale in ring-tailed coatis (Nasua nasua) of the Tiete Ecological Park, São Paulo, Brazil. Pesqui Veterinária Bras. **29**, 959–962 (2009). https://doi.org/10.1590/s0100-736x2009001200001
4. Pesenti, T.C., Mascarenhas, C.S., Krüger, C., et al.: Dioctophyma renale (Goeze, 1782) Collet - Meygret, 1802 (Dioctophymatidae) in Galictis cuja (Molina, 1782) (Mustelidae) de Rio Grande do Sul, Brazil. Neotrop Helminthol. **6**, 301–305 (2012)
5. Duarte, J., Costa, A.M.B., Katagiri, S., et al.: Parasitism by Dioctophyme renale (Goeze, 1782) in maned wolf (Chrysocyon brachyurus), Brazil. Veterinária e Zootec. **20**, 52–56 (2013)
6. Perera, S.C., Capella, G.d.A., Pinto, N.B., et al.: First isolation of Dioctophyme renale eggs from an urban environment and identification of those from animal urine. Brazilian J Vet Parasitol. **26**, 89–91 (2017). https://doi.org/10.1590/s1984-29612016064
7. Trindade, M.A.C., De Macedo, M.R.P., Muller, G.: Dioctophyme renale (Nematoda: Dioctophymatidae)

in Leopardus geoffroyi (Carnivora: Felidae) in the Neotropical region. Brazilian J Vet Parasitol. **27**, 223–225 (2018). https://doi.org/10.1590/S1984-29612017079
8. Pedrassani, D., Nascimento, A.A.: Verme gigante renal. Parasite Giant Renal. **110**, 30–37 (2015)
9. Li, G., Liu, C., Li, F., et al.: Fatal bilateral dioctophymatosis. J. Parasitol. **96**, 1152–1154 (2010). https://doi.org/10.1645/ge-2132.1
10. Sapin, C.F., Silva-Mariano, L.C., Grecco-Corrêa, L., et al.: Dioctofimatose renal bilateral e disseminada em cão. Pesqui. Vet. Bras. **37**, 1499–1504 (2017). https://doi.org/10.1590/S0100-736X2017001200022
11. Regalin, B.D.d.C., Tochedo, R., Colodel, M.M., et al.: Dioctophyma renale em testículo de cão. Parasiten des Menschen. **44**, 1–4 (2016). https://doi.org/10.1007/978-3-642-85397-5_28
12. Perera, S.C., Rappeti, J.C.S., Milech, V., et al.: Eliminação de Dioctophyme renale pela urina em canino com dioctofimatose em rim esquerdo e cavidade abdominal – Primeiro relato no Rio Grande do Su. Arq Bras Med Vet e Zootec. **69**, 618–622 (2017)
13. Mace, T.F., Anderson, R.C.: Development of the giant kidney worm, Dioctophyma renale (Goeze, 1782) (Nematoda: Dioctophymatoidea). Can. J. Zool. **53**, 1552–1568 (1975)
14. Nakagawa, T.L.D.R., Bracarense, A.P.F.R.L., dos Reis, A.C.F., et al.: Giant kidney worm (Dioctophyma renale) infections in dogs from Northern Paraná, Brazil. Vet. Parasitol. **145**, 366–370 (2007). https://doi.org/10.1016/j.vetpar.2006.10.027
15. Stainki, D.R., Pedrozo, J.C.S.R., Gaspar, L.F.J., et al.: Urethral obstruction by Dioctophyma renale in puppy. Comp. Clin. Pathol. **20**, 535–537 (2011). https://doi.org/10.1007/s00580-010-1169-0
16. da Veiga, C.C.P., de Oliveira, P.C., AMR, F., et al.: Dioctofimose em útero gravídico em cão – Relato de caso. Revista Brasileira de Medicina Veterinária. **34**, 188–191 (2012)
17. Figueiredo, M.A.P., da Silva, D.F., Manrique, W.G., de Sousa, A.A.R.: Ciclo errático de Dioctophyme renale: relato de dois casos. Orinoquia. **17**, 96–101 (2013). https://doi.org/10.22579/20112629.54
18. Russo, Z.H., Rizzardini, E.S., Nuñez, C.d.S., Monte, F.N.: Dioctophyma renale in dogs (Canis familiaris) in Uruguay. Neotrop Helminthol. **8**, 123–130 (2014)
19. Silveira, C.S., Diefenbach, A., Mistieri, M.L., et al.: Dioctophyma renale em 28 cães: aspectos clinicopatológicos e ultrassonográficos. Pesqui. Vet. Bras. **35**, 899–905 (2015). https://doi.org/10.1590/s0100-736x2015001100005
20. Bach, F.S., Klaumann, P.R., Montiani-Ferreira, F.: Paraparesis secondary to erratic migration of Dioctophyma renale in a dog. Ciência Rural. **46**, 885–888 (2016). https://doi.org/10.1590/0103-8478cr20151219
21. Sardjono, T.W., Purnomo, B.B., Iskandar, A., Gunawan, A.: Dioctophymatosis renalis in humans: first case report from Indonesia. Proc 3rd ASEAN Congr Trop Med Parasitol. **3**, 90–93 (2008)
22. Measures, L.N., Anderson, R.C.: Centrarchid fish as paratenic hosts of the giant kidney worm, Dioctophyma renale (Goeze, 1782), in Ontario, Canada. J. Wildl. Dis. **21**, 11–19 (1985)
23. Anderson, R.C.: Nematode parasites of vertebrates: their development and transmission, 2nd edn. CAB International, Wallingford (2000)
24. Lisboa, A.: Estrongilose renal humana. Bras. Med., 23–24 (1945)
25. Fernando, S.S.E.: The giant kidney worm (Dioctophyma renale) infection in man in Australia. Am. J. Surg. Pathol. **7**, 281–284 (1983)
26. Gutierrez, Y., Cohen, M., Machicao, C.N.: Dioctophyme larva in the subcutaneous tissues of a woman in Ohio. Am. J. Surg. Pathol. **13**, 800–802 (1989)
27. Katafigiotis, I., Fragkiadis, E., Pournaras, C., et al.: A rare case of a 39 year old male with a parasite called Dioctophyma renale mimicking renal cancer at the computed tomography of the right kidney. A case report. Parasitol. Int. **62**, 459–460 (2013). https://doi.org/10.1016/j.parint.2013.06.007
28. Abdel-Hakeem, S.S., Abdel-Samiee, M.A.: Case study: Dioctophyma renale infection in mice, incidental finding during experimental studies. Egypt Acad. J. Biol. Sci. **10**, 83–91 (2018). https://doi.org/10.21l3/gselements.7.5.313
29. Eiras, J., Zhu, X, Yurlova, N., et al.: Dioctophyme renale (Goeze, 1782) (Nematoda, Dioctophymidae) parasitic in mammals other than humans: a comprehensive review. Parasitol. Int., 81, 1-31 (2021). https://doi.org/10.1016/j.parint.2020.102269
30. Ferreira, V.L., Medeiros, F.P., July, J.R., Raso, T.F.: Dioctophyma renale in a dog: clinical diagnosis and surgical treatment. Vet. Parasitol. **168**, 151–155 (2010). https://doi.org/10.1016/j.vetpar.2009.10.013
31. Vulcani, V.A.S., Franzo, V.S., de Araújo, D.P., et al.: Dioctophyma renale em lobo-guará na região geoeconômica de Jataí, GO, Brasil - Relato de caso. Rev Bras Med Veterinária. **37**, 149–152 (2015)
32. Lima, V.F.S., Ramos, R.A.N., Lepold, R., et al.: Use of the FLOTAC technique to diagnosing parasites of the urinary tract of dogs. Parasitol. Res. **115**, 1737–1739 (2016). https://doi.org/10.1007/s00436-016-4950-4
33. Faria, B.M., Pereira, K.d.F., Leão, A.P., et al.: Diagnosis of Dioctophyma renale using ultrasonography and microscopy – a case report. Veterinária Em Foco. **14**, 48–53 (2017)
34. Pedrassani, D., Wendt, H., Rennau, E.A., et al.: Dioctophyme renale Goeze, 1782 in a cat with a supernumerary kidney. Brazilian J Vet Parasitol. **23**, 109–111 (2014). https://doi.org/10.1590/s1984-29612014018
35. Pedrassani, D., Nascimento, A.A.d., André, M.R., Machado, R.Z.: Dioctophyme renale: prevalence and risk factors of parasitism in dogs of São Cristóvão district, Três Barras county, Santa Catarina state,

Brazil. Brazilian J Vet Parasitol. **26**, 39–46 (2017a). https://doi.org/10.1590/s1984-29612017004
36. Vieira, F.M., Muniz-Pereira, L.C., de Souza-Lima, S., et al.: Parasitic nematodes of three species of wild carnivore mammals from Atlantic forest in the state of Minas Gerais, Brazil. Rev Mex Biodivers. **88**, 801–806 (2017). https://doi.org/10.1016/j.rmb.2017.10.033
37. de Freitas, D.M., Maria, B.P., Vasconcelos, B.M.A., et al.: Nefrectomia unilateral em um cão parasitado por Dioctophyma renale: relato de caso. PUBVET. **12**, 1–7 (2018). https://doi.org/10.31533/pubvet.v12n9a178.1-7
38. Pizzinatto, F.D., Freschi, N., Sônego, D.A., et al.: Parasitism by Dioctophyma renale in a dog: clinical and surgical aspects. Acta Sci. Vet. **47**, 1–7 (2019). https://doi.org/10.22456/1679-9216.93924
39. Ribeiro, C.T., Verocal, G.G., Tavares, L.E.R.: Dioctophyme renale (Nematoda, Dioctophymatidae) infection in the Crab-eating Fox (Cerdocyon thous) from Brazil. J. Wildl. Dis. **45**, 248–250 (2009). https://doi.org/10.7589/0090-3558-45.1.248
40. Verocai, G.G., Measures, L.N., Azevedo, F.D., et al.: Dioctophyme renale (Goeze, 1782) in the abdominal cavity of a domestic cat from Brazil. Vet. Parasitol. **161**, 342–344 (2009). https://doi.org/10.1016/j.vetpar.2009.01.032
41. Zabott, M.V., Pinto, S.B., Viott, A.M., et al.: Ocorrência de Dioctophyma renale em Galictis cuja. Pesqui Veterinária Bras. **32**, 786–788 (2012). https://doi.org/10.1590/S0100-736X2012000800018
42. Pedrassani, D., Worm, M., Drechmer, J., Santos, M.C.I.: Lesser Grison (Galictis cuja Molina, 1782) as host of Dioctophyme renale Goeze, 1782. Arq Inst Biol (Sao Paulo). **84**, 1–4 (2017b). https://doi.org/10.1590/1808-1657000312016
43. da Silva, C.C., Cavalcanti, G.A.d.O., Braga, F.d.V.A., et al.: Dioctophyme renale parasitism in domestic cat in Rio Grande do Sul, Brazil – first reported case. Semin Ciências Agrárias. **38**, 2215–2220 (2017). https://doi.org/10.5433/1679-0359.2017v38n4p2215
44. Rocco, F.d.S., Sanavria, A., Ribeiro, B.C.C., et al.: Relato de casos : dioctofimose em cães provenientes do Município de Angra dos Reis, Estado do Rio de Janeiro. An da XI Jorn Iniciação Científica da UFRRJ. **11**, 139–140 (2001)
45. Kano, F.S., Shimada, M.T., Suzuki, S.N., et al.: Dioctophymosis occurrence in two dogs in Guarapuava city – Parana state. Semin Ciências Agrárias. **24**, 177–180 (2003)
46. Pereira, B.J., Girardelli, G.L., Trivilin, L.O., et al.: Ocorrência de dioctofimose em cães do município de Cachoeiro do Itapemirim, Espírito Santo, Brasil, no período de maio a dezembro de 2014. Rev. Bras. Parasitol. Vet. **15**, 123–125 (2006)
47. Ishizaki, M.N., Imbeloni, A.A., Muniz, J.A.P.C., et al.: Dioctophyma renale (Goeze, 1782) in the abdominal cavity of a capuchin monkey (Cebus apella), Brazil. Vet. Parasitol. **173**, 340–343 (2010). https://doi.org/10.1016/j.vetpar.2010.07.003
48. Cansi, E.R., Bonorino, R., Mustafa, V.S., Guedes, K.M.R.: Multiple parasitism in wild maned wolf (Chrysocyon brachyurus, Mammalia: Canidae) in Central Brazil. Comp. Clin. Pathol. **21**, 489–493 (2012). https://doi.org/10.1007/s00580-012-1513-7
49. Eicke, B., Zanetti, K., Reginaldo, A.S., et al.: Dioctophyma renale in a patient living in Mato Grosso do Sul. Acta Vet Bras. **8**, 282–285 (2014). https://doi.org/10.21708/avb.2014.8.4.5175
50. Kommers, G.D., Ilha, M.R.d.S., Barros, C.S.L.d.: Dioctofimose em cães: 16 casos. Ciência Rural. **29**, 517–522 (1999). https://doi.org/10.1590/s0103-84781999000300023
51. Brun, M.V., Beck, C.A.d.C., Mariano, M.B., et al.: Nefrectomia laparoscópica em cão parasitado por Dioctophyma renale – relato de caso. Arq Ciências Veterinárias e Zool da UNIPAR. **5**, 145–152 (2002)
52. Secchi, P., Valle, S.d.F., Brun, M.V., et al.: Nefrectomia videolaparoscópica para tratamento da dioctofimose em um cão. Acta Sci. Vet. **38**, 85–89 (2010). https://doi.org/10.22456/1679-9216.16550
53. Rappeti, J.C.d.S., Mascarenhas, C.S., Perera, S.C., et al.: Dioctophyme renale (Nematoda: Enoplida) in domestic dogs and cats in the extreme South of Brazil. Brazilian J Vet Parasitol. **26**, 119–121 (2017). https://doi.org/10.1590/s1984-29612016072
54. Echenique, J.V.Z., Soares, M.P., Mascarenhas, C.S., et al.: Lontra longicaudis infected with canine parvovirus and parasitized by Dioctophyma renale. Pesqui Veterinária Bras. **38**, 1844–1848 (2018). https://doi.org/10.1590/1678-5150-pvb-5744
55. Perera, S.C., Milech, V., Wulfing, C.T., et al.: Sobrevivência do Dioctophyma renale na cavidade abdominal do cão. An do 21° Congr Iniciação Científica da Univ Fed Pelotas (2012)
56. Trindade, M.A.C., Portela, P.R., Mascarenhas, C.S., et al.: Larvas de Dioctophyme renale (Goeze, 1782) (Nematoda: Enoplida) em Procyon cancrivorus (Cuvier, 1798) (Carnivora: Procyonidae) no sul do Brasil. In: Anais do XXI Encontro de Pós-Graduação da Universidade Federal de Pelotas (2019)
57. Mascarenhas, C.S., Müller, G.: Third-stage larvae of the enoplid nematode Dioctophyme renale (Goeze, 1782) in the freshwater turtle Trachemys dorbigni from southern Brazil. J. Helminthol., 1–6 (2014). https://doi.org/10.1017/S0022149X14000364
58. Mascarenhas, C.S., Henzel, A.B.D., Corrêa, F., et al.: Third-stage larvae of Dioctophyme renale (Goeze, 1782) (Nematoda: Enoplida) in Hoplosternum littorale (Hancock, 1828) (Siluriformes: Callichthyidae) from Southern Brazil. Neotrop Helminthol. **10**, 135–138 (2016)
59. Mascarenhas, C.S., Henzel, A.B.D., Müller, G.: Larvae of Dioctophyme renale (Goeze, 1782) (Nematoda: Enoplida) and Contracaecum sp. (Nematoda: Anisakidae) in freshwater turtles (Testudines: Chelidae) from Southern Brazil. Neotrop Helminthol. **11**, 61–67 (2017)

60. Mascarenhas, C.S., Pereira, J.V., Müller, G.: Occurrence of Dioctophyme renale larvae (Goeze, 1782) (Nematoda: Enoplida) in a new host from southern Brazil. Brazilian J Vet Parasitol. **27**, 609–613 (2018). https://doi.org/10.1590/s1984-296120180060
61. Mascarenhas, C.S., Müller, G., de Macedo, M.R.P., et al.: The role of freshwater fish in the life cycle of Dioctophyme renale in Southern Brazil. Vet Parasitol Reg Stud Reports. **16**, 1–7 (2019). https://doi.org/10.1016/j.vprsr.2019.100274
62. Pedrozo, R., Bazán, A., Cuevas, D., Godoy, Y.: Valores séricos de urea, creatinina y densidad urinaria en perros com Dioctophyma renale en la ciudad de Caazapá, Paraguay, y su relación com factores de riesgo. Compend Ciencias Vet. **3**, 33–38 (2013)
63. Radman, N.E., Gamboa, M.I., Butti, M.J., et al.: Occurrence of dioctophymosis in canines within a riparian zone of the Río de La Plata watercourse, in Ensenada, Buenos Aires Province, Argentina. Vet Parasitol Reg Stud Reports. **10**, 43–50 (2017). https://doi.org/10.1016/j.vprsr.2017.07.007
64. Flórez, Á.A., Russo, J., Uribe, N.: First report of Dioctophyma renale (Nematoda, Dioctophymatidae) in Colombia. Biomedica. **38**, 13–18 (2018). https://doi.org/10.7705/BIOMEDICA.V38I4.4042
65. Unruh, D.H.A., King, J.E., Eaton, R.D.P., Allen, J.R.: Parasites of dogs from Indian settlements in northwestern Canada: a survey with public health implications. Can. J. Comp. Med. **37**, 25–32 (1973)
66. Paras, K.L., Miller, L., Verocai, G.G.: Ectopic infection by Dioctophyme renale in a dog from Georgia, USA, and a review of cases of ectopic dioctophymosis in companion animals in the Americas. Vet Parasitol Reg Stud Reports. **14**, 111–116 (2018). https://doi.org/10.1016/j.vprsr.2018.09.008
67. Gargili, A., Firat, I., Toparlak, M., Çetinkaya, H.: First case report of Dioctophyme renale (Goeze, 1782) in a dog in Üstanbul, Turkey. Turkish J Vet Anim Sci. **26**, 1189–1191 (2002)
68. Safarov, A.A., Akramova, F.D., Shakarbaev, U.A., Azimov, D.A.: Parasitic fauna of domestic dog (Canis familiaris Dom.) of the modern metropolis of Tashkent. Russ J Parasitol. **12**, 41–49 (2018). https://doi.org/10.31016/1998-8435-2018-12-4-41-49
69. Zolhavarieh, S.M., Norian, A., Yavari, M.: Dioctophyma renale (Goeze, 1782) infection in a domestic dog from Hamedan, Western Iran. Iran. J. Parasitol. **11**, 131–135 (2016)
70. Shahbazi, P., Shahbazfar, A.A., Norouzi, R.: A case report of Dioctophyma renale in the peritoneal cavity of a stray dog. J Zoonotic Dis. **2**, 43–47 (2017)
71. Wang, C.R., Qiu, J.H., Zhao, J.P., et al.: Prevalence of helminthes in adult dogs in Heilongjiang Province, the People's republic of China. Parasitol. Res. **99**, 627–630 (2006). https://doi.org/10.1007/s00436-006-0219-7
72. McLeod, J.A.: Dioctophyma renale infections in Manitoba. Can. J. Zool. **45**, 505–508 (1967). https://doi.org/10.1139/z67-064
73. Burgos, L., Acosta, R.M., Fonrouge, R.D., et al.: Prevalence of a zoonotic parasite, Dioctophyma renale (Goeze, 1782), among male canines in a wild riverside area of la plata river, province of Buenos Aires, Republic of Argentina. Rev Patol Trop. **43**, 420–426 (2014). https://doi.org/10.5216/rpt.v43i4.33603
74. Al-Obaidi, Q.T.: Prevalence of internal helminthes in stray cats (Felis catus) in Mosul City, Mosul-Iraq. J. Anim. Vet. Adv. **11**, 2732–2736 (2012)
75. Butti, M.J., Gamboa, M.I., Terminiello, J.D., et al.: Dioctophyme renale in a domestic cat (Felis catus): renal location and nephrectomy. Vet Parasitol Reg Stud Reports. **18**, 1–5 (2019). https://doi.org/10.1016/j.vprsr.2019.100339
76. Sapin, C.d.F., Silva-Mariano, L.C., Piovesan, A.D., et al.: Anatomopathologic study of kidneys parasitized by Dioctophyme renale in dogs. Acta Sci. Vet. **45**, 1–7 (2017). https://doi.org/10.22456/1679-9216.79788
77. Leite, L.C., Círio, S.M., Diniz, J.M.F., et al.: Anatomopathologic lesions found in Dioctophyma renale (Goeze, 1782) infections in domestic dogs (Canis familiaris, LINNAEUS, 1758). Arch Vet. **10**, 95–101 (2005)
78. Monteiro, S.G., Sallis, E.S.V., Stainki, D.R.: Natural infection by thirty four helminthes of the Dioctophyma renale (Goeze, 1782) species in a dog. Rev da FZVA. **9**, 95–99 (2002)
79. Alves, G.C., Silva, D., Neves, M.F.: Dioctophyma renale: o parasita gigante do rim. Rev Científica Eletrônica Med Veterinária (2007)
80. Kumar, V., Vercruysse, J., Vandesteene, R.: Studies on two cases of Dioctophyma renale (Goeze, 1782) infection in Chrysocyon brachyurus (Illiger). Acta Zool. Pathol. Antverp. **56**, 83–98 (1972)
81. Hernández-Camacho, N., Pineda-López, R., López-González, C.A., Jones, R.W.: Nematodes parasites of the gray fox (Urocyon cinereoargenteus Schreber, 1775) in the seasonally dry tropical highlands of Central Mexico. Parasitol. Res. **108**, 1425–1429 (2011). https://doi.org/10.1007/s00436-010-2191-5
82. Holmes, J.C., Podesta, R.: The helminths of wolves and coyotes from the forested regions of Alberta. Can. J. Zool. **46**, 1193–1204 (1968). https://doi.org/10.1139/z68-169
83. Seville, R.S., Addison, E.M.: Nongastrointestinal helminths in marten (Martes americana) from Ontario, Canada. J. Wildl. Dis. **31**, 529–533 (1995). https://doi.org/10.7589/0090-3558-31.4.529
84. Harkema, R., Miller, G.C.: Helminth parasites of the Raccoon Procyon lotor in the Southeastern United States. J. Parasitol. **50**, 60–66 (1964)
85. Mech, L.D., Tracy, S.P.: Prevalence of giant kidney worm (Dioctophyma renale) in wild

mink (Mustela vison) in Minnesota. Am. Midl. Nat. **145**, 206–209 (2001) https://doi.org/10.1674/0003-0031(2001)145[0206:pogkwd]2.0.co;2
86. Hoffman, V., Nolan, T.J., Schoelkopf, R.: First report of the giant kidney worm (Dioctophyme renale) in a harbor seal (Phoca vitulina). J. Parasitol. **90**, 659–660 (2004). https://doi.org/10.1645/ge-198r
87. Loukmas, J.J., Mayack, D.T., Richmond, M.E.: Giant kidney worms in mink from New York: prevalence, distribution, and health implications. Northeast. Nat. **17**, 211–222 (2010). https://doi.org/10.1656/045.017.0204
88. Kauffman, J.C., Harle, K.K., Swallow, M.J., Huffman, J.E.: Case report: occurrence of the giant kidney worm (Dioctophyma renale) in long-tailed weasels (Mustela frenata) from Pennsylvania. J Pennsylvania Acad Sci. **87**, 50–52 (2013)
89. Hajialilo, E., Mobedi, I., Masoud, J., et al.: Dioctophyme renale in Vulpes vulpes from the Caspian Sea littoral of Iran. Iran. J. Public Health. **44**, 698–700 (2015)
90. Tokiwa, T., Harunari, T., Tanikawa, T., et al.: Dioctophyme renale (Nematoda: Dioctophymatoidea) in the abdominal cavity of Rattus norvegicus in Japan. Parasitol. Int. **60**, 324–326 (2011). https://doi.org/10.1016/j.parint.2011.03.003
91. Corrêa, P., Bueno, C., Vieira, F.M., Muniz-Pereira, L.C.: Helminth parasites of Galictis cuja (Carnivora, Mustelidae), from localities in the Atlantic forest of Brazil. Brazilian J Vet Parasitol. **25**, 407–413 (2016). https://doi.org/10.1590/S1984-29612016077
92. Pedrassani, D., Hoppe, E.G.L., Tebaldi, J.H., Nascimento, A.A.d.: Chaunus ictericus (Spix, 1824) as paratenic host of the giant kidney worm Dioctophyme renale (Goeze, 1782) (Nematoda: Enoplida) in São Cristóvão district, Três Barras county, Santa Catarina state, Brazil. Vet. Parasitol. **165**, 74–77 (2009). https://doi.org/10.1016/j.vetpar.2009.07.017
93. Urano, Z., Hasegawa, H., Katsumata, T., et al.: Dioctophymatid nematode larva found from human skin with creeping eruption. J. Parasitol. **87**, 462–465 (2001). https://doi.org/10.2307/3285077
94. Park, H.Y., Seo, J.W., Lee, B.H., et al.: Simultaneous occurrence of malignant fibrous histiocytoma of the ureter and Dioctophyma renale infection: a case report. J Korean Soc Radiol. **68**, 411–415 (2013). https://doi.org/10.3348/jksr.2013.68.5.411
95. Agrawal, R., Kumar, P., Mishra, P.P., et al.: Dioctophyma renale: a chance finding on bladder catheterisation of a pregnant female. Indian J Med Case Reports. **3**, 70–72 (2014)
96. Tokiwa, T., Ueda, W., Takatsuka, S., et al.: The first genetically confirmed case of Dioctophyme renale (Nematoda: Dioctophymatida) in a patient with a subcutaneous nodule. Parasitol. Int. **63**, 143–147 (2014). https://doi.org/10.1016/j.parint.2013.09.015
97. Norouzi, R., Manochehri, A., Hanifi, M.: A case report of human infection with Dioctophyma renale from Iran. Urol. J. **14**, 3043–3045 (2017). https://doi.org/10.22037/uj.v14i2.3722
98. Sun, T., Turnbull, A., Lieberman, P.H., Sternberg, S.S.: Giant kidney worm (Dioctophyma renale) infection mimicking retroperitoneal neoplasm. Am. J. Surg. Pathol. **10**, 508–512 (1986)
99. Ignjatovic, I., Stojkovic, I., Kutlesic, C., Tasic, S.: Infestation of the human kidney with Dioctophyma renale. Urol. Int. **70**, 70–73 (2003). https://doi.org/10.1159/000067695
100. Gu, Y., Li, G., Zhang, J., Zhang, Y.: Dioctophyma renale infection masquerading as a malignancy. Kidney Int. **82**, 1342 (2012). https://doi.org/10.1038/ki.2012.315
101. Chauhan, S., Kaval, S., Tewari, S.: Dioctophymiasis: a rare case report. J. Clin. Diagn. Res. **10**, 1–2 (2016). https://doi.org/10.7860/JCDR/2016/17394.7305
102. Yang, J., Li, P., Su, C., et al.: Worms expelled with the urine from a bosniak cyst III of the left kidney. Urology. **93**, e5 (2016). https://doi.org/10.1016/j.urology.2016.02.024
103. Kuehn, J., Lombardo, L., Janda, W.M., Hollowell, C.M.P.: Giant kidney worms in a patient with renal cell carcinoma. BMJ Case Rep., 1–4 (2016). https://doi.org/10.1136/bcr-2015-212118
104. Acha, P.N., Skyfres, B.: Zoonoses and communicable diseases common to man and animals (Volume III – Parasitosis), 3rd edn. Pan American Health Organization, Washington (2003)
105. Rahal, S.C., Mamprim, M.J., Oliveira, H.S., et al.: Ultrasonographic, computed tomographic, and operative findings in dogs infested with giant kidney worms (Dioctophyme renale). J. Am. Vet. Med. Assoc. **244**, 555–558 (2014). https://doi.org/10.2460/javma.244.5.555
106. Lopes, S.T.d.A., Veiga, A.: Urinálise. In: González, F.H.D., da Silva, S.C. (eds.) Patologia Clínica Veterinária: Texto Introdutório, pp. 107–139. Universidade Federal do Rio Grande do Sul, Porto Alegre (2008)
107. Stockham, S.L., Scott, M.A.: Fundamentals of Veterinary Clinical Pathology, 2nd edn. Blackwell Publishing, Iowa (2008)
108. Pedrassani, D., Hoppe, E.G.L., Avancini, N., Nascimento, A.A.d.: Morphology of eggs of Dioctophyme renale Goeze, 1782 (Nematoda: Dioctophymatidae) and infuences of temperature on development of first-stage larvae in the eggs. Rev. Bras. Parasitol. Vet. **18**, 15–19 (2009). https://doi.org/10.4322/rbpv.01801003
109. Pedrassani, D.: Aspectos morfológicos, imunológicos e epidemiológicos do Dioctophyme renale em cães no Distrito de São Cristóvão, Três Barras, Santa Catarina. Universidade Paulista "Julio de Mesquita Filho" (2009)
110. Sink, C.A., Feldman, B.F.: Laboratory Urinalysis and Hematology for the Small Animal

Practitioner, 1st edn. Teton NewMedia, Jackson, WY (2004)

111. Sink, C.A., Weinstein, N.M.: Practical Veterinary Urinalysis, 1st edn. Wiley-Blackwell, Chichester (2012)
112. Lopes, S.T.d.A., Biondo, A.W., Santos, A.P.d.: Manual de Patologia Clínica Veterinária, 3rd edn. UFSM/Departamento de Clínica de Pequenos Animais, Santa Maria (2007)
113. Urqhuart, G.M., Armour, J., Duncan, J.L., et al.: Parasitologia Veterinária, 2nd edn. Guanabara Koogan S.A, Rio de Janeiro (1998)
114. Bédard, C., Desnoyers, M., Lavallée, M.-C., Poirier, D.: Capillaria in the bladder of an adult cat. Can. Vet. J. **43**, 973–974 (2002)
115. Pedrassani, D., Nascimento, A.A.d., André, M.R., Machado, R.Z.: Improvement of an enzyme immunosorbent assay for detecting antibodies against Dioctophyma renale. Vet. Parasitol. **212**, 435–438 (2015). https://doi.org/10.1016/j.vetpar.2015.08.007
116. Giorello, A.N., Kennedy, M.W., Butti, M.J., et al.: Identification and characterization of the major pseudocoelomic proteins of the giant kidney worm, Dioctophyme renale. Parasit. Vectors. **10**, 1–11 (2017). https://doi.org/10.1186/s13071-017-2388-x
117. Capella, G. de A., Pinto, N.B., Moura, M.Q. de., et al.: Imunodiagnóstico de dioctofimatose em cães. An do XXI Encontro Pós-Graduação da Univ Fed Pelotas (2019)

Index

H. M. Baptista Carreira dos Santos (ed.), *Translational Urinomics*, Advances in Experimental Medicine and Biology 1306, https://doi.org/10.1007/978-3-030-63908-2

The manufacturer's authorised representative in the EU is Springer Nature Customer Service Centre GmbH, Europaplatz 3, 69115 Heidelberg, Germany. If you have any concerns regarding our products, please contact ProductSafety@springernature.com

Printed and bound by CPI Group (UK) Ltd, Croydon, CR0 4YY
30/09/2026
02225665-0017